FUNDAMENTALS OF SOLID-STATE ELECTRONICS

Solution Manual

FUNDAMENTALS OF SOLID-STATE ELECTRONICS

Solution Manual

Chih-Tang Sah

The Robert C. Pittman Eminent Scholar Chair
Graduate Research Professor of Electrical and Computer Engineering
University of Florida
Gainesville, Florida
U.S.A.

Published by

World Scientific Publishing Co. Pte. Ltd.

P O Box 128, Farrer Road, Singapore 912805

USA office: Suite 1B, 1060 Main Street, River Edge, NJ 07661

UK office: 57 Shelton Street, Covent Garden, London WC2H 9HE

Library of Congress Cataloging-in-Publication Data

Sah, Chih-Tang.

Fundamentals of solid-state electronics - solution manual/Chih-Tang Sah.

p. cm.

Includes bibliographical references.

ISBN 9810228813 (pbk)

1. Solid state electronics -- Problems, exercises, etc. I. Title.

TK7871.85.S23 1996

621.381'076--dc20 96-29406

CIP

British Library Cataloguing-in-Publication Data

A catalogue record for this book is available from the British Library.

Printed in Singapore by Uto-Print

Preface

This is the **solution manual** for the textbook, **Fundamentals of Solid-State Electronics** (first edition, October 1991) (**FSSE**). It contains the solution of more than 100 FSSE problems assigned by this author and his colleagues in a one-semester junior-level introductory core course on solid-state devices from 1989 to 1996 at the University of Florida. A copy of the solutions was given to the students with their graded solutions.

This solution manual can be acquired by both students and instructors so that the students can study the methods of solution and acquire a deeper understanding of the underlying physics from the descriptions and essay solutions. Thus, it is not only a **teaching guide** for the instructors but also a **learning aid** for the students.

An extensive appendix on **transistor reliability** is added because of current manufacturing concerns on the operation time-to-failure of transistors. It contains descriptions of the fundamental causes of transistor instability and the solution of fundamental and applied problems. Four subjects are covered: interface trap, oxide trap, acceptor hydrogenation, and electromigration. The materials were taught in a senior transistor device and circuit reliability course by Toshikazu Nishida and this author and in a third-year advanced graduate course on semiconductor material and device physics by this author during 1988-1996 at the University of Florida.

I would like to acknowledge Professors Toshikazu Nishida and Arnost Neugroschel for using FSSE and FSSE-Study Guide to teach the junior and advanced courses, and providing initial versions of some of the solutions. Some figures and data were taken from the doctoral thesis and journal articles of our first four graduate students at the University of Florida, Scott Thompson, Yi Lu, Jack Kavalieros and Michael Carroll. Unpublished data from current graduate students Michael Han and Kurt Pfaff were also consulted or used. Proofreading of the final drafts was assisted by Mike Han, Derek Martin, Jin Cai, and Professors Nishida and Neugroschel. Special appreciation is due Ms. Barbara Aman, May Look, and Yanan Fan (World Scientific, Singapore) for publication supports. I also thank Linda Sah for assistance and encouragement.

Chih-Tang Sah
Gainesville, Florida
July 4, 1996

FUNDAMENTALS OF SOLID-STATE ELECTRONICS

Solution Manual

CONTENTS

Chapters and Appendix

Chapters and Appendix

Chapter 1
ELECTRONS, BONDS, BANDS, AND HOLES

OBJECTIVES

*** Understand two fundamental electrical concepts**

– Electrons

– Holes

*** Understand two semiconductor materials models**

– Bond Model (or Valence Bond Model)

– Band Model (or Energy Band Model)

P100.1 Describe the two main physical causes that make classical Newtonian and classical statistical mechanics inadequate to describe the motion of electrons and atoms in a solid. What hypotheses, laws, principles, models, or methodologies are introduced in quantum mechanics and quantum statistical mechanics to overcome these limitations?

The explanations are given on pages 2-3 and 31-35 of FSSE. This question should be asked again and its answers thought through again at the end of chapter 3 when the students have learned the details of the fundamental physics concepts underlying the answers to this question. The detailed answers given below explain the fundamental reasons which the instructor can repeat in a class room discussion and the students can use as a review focus. The keyword-answers can be succinctly given in a few sentences which are underlined and could be adequate to receive full credit for the problem, provided the student understands the physics underlying the keywords as explained in the FSSE and summarized below. This presentation format (underlining) of the solutions is followed for some of the discussion and essay problems in this Solution Manual.

The two main physical causes are the large particle number (about 10^{23} particles in one cubic centimeter) and small interparticle distance (about one angstrom or 10^{-8} cm) in a solid. These combined, translate to high particle density which is a consequence rather than a cause.

The small interparticle distance means the force between two particles must be taken into account because for the charged particles (electrons and ionized atomic nuclei in a solid) the interparticle force is the long-range, 1/r dependent Coulomb force whose range is large compared with the interparticle spacing. (Actually, the Coulomb force has an 'infinite' range so the Coulombic interparticle force must be taken into account to analyze the particle motion of a many-particle system at all interparticle spacings, but the smaller spacings make the force more important. The particles cannot be treated as hard sphere with an infinite or large repulsive interparticle force at their radii and zero force when separated.)

The long-range interparticle force means that one must simultaneously solve the many Newton's equations of force and mass (10^{23} equations in one cm^3 of solid), one equation from each particle, which are coupled by the interparticle force. This coupling through the interparticle force that affects the trajectory of each particle is known as **scattering**. The small interparticle distance and the large number of particles mean that each particle experiences very many scattering events or deflections of their trajectories in an experimenter's observation time, about 10^{14} per second. It is obviously impossible to solve such a system of 10^{23} equations using any known computation equipment and methods. On the contrary, the motion (position and velocity at any given time) of celestial bodies are indeed obtained by solving simultaneously the Newton's equations, one for each body, which are coupled through the interbody gravitation force, but the solutions are readily obtained only for a partitioned or local celestial system that contains a few interacting bodies. For examples, earth, moon and sun is a three-body system; meteor composed of many small bodies clustered together is treated as one body when solving for its orbit around the sun; and similarly, the many electrons and atoms composing the earth (or sun or moon) are lumped together in one earth body in the three-body earth-moon-sun system.

A more practical reason in solid, unlike the celestial bodies, is that we cannot experimentally measure the motion (position and velocity) of each of the 10^{23} particles (electrons and atoms or nuclei) in a solid even with the sharpest electron or optical microscope to 'see' them or the most sensitive charge detection meter to 'feel' them, for two reasons: (i) electron or photon which is scattered off the particle during the process of 'seeing' the particle will have altered the position and velocity of the particle during the scattering, and (ii) the electrical properties of the solid we can measure with a meter or an oscilloscope are averaged properties over many particles and the motion of onc particle among the 10^{23} is not relevant.

Thus, the classical mechanics of Newton is inadequate to analyze the many particle problem of a solid because there are too many equations, one for each particle, to be solved simultaneously and its solution, even if obtainable, is not what we observe and measure because the measuring particle (an x-ray photon or electron-microscope keV-electron) would have altered both the position and velocity of the particle being measured when scattered off the particle during the measurement.

Thus, there are two parts of the solid property problem to solve: (i) to describe the motion of a single particle whose position and velocity would have to be described in such a mathematical way in order to take into account that its motion would be changed by the measuring particle scattering off the measured particle, and (ii) to describe the motion of many interacting particles whose average properties are observed and measured in the laboratory. Both require the use of the probability concept.

First, we need a theory to describe the probability of finding a particle at a particular position (or in a finite volume element $\Delta x \Delta y \Delta z$ located at x, y, and z) with a particular velocity (or within a velocity range, v_x to v_x+dv_x, v_y to v_y+dv_y, and v_z to v_z+dv_z) and at a given time (or in a time interval, t to t+dt). A differential equation is the preference because it allows us to obtain the solutions for different initial (time) and boundary (geometry) conditions and for different force laws. This is the Schrödinger partial differential equation proposed in 1925 whose solution is a wave-like function which is defined as the probability amplitude of finding the particle in ($\Delta x, \Delta y, \Delta z, \Delta t$) at (x,y,z,t). It is based on the wave-like properties of particles, hypothesized by deBroglie in 1924, and the particle-like properties of light or electromagnetic waves, hypothesized by Planck in 1900.

The single-particle solution of the Schrödinger equation must take up one further step to deal with the many particles in a solid since it is the average properties of the many particles we are observing and measuring in the laboratory or everyday life. The many particles means that the theory must be statistical in nature to allow calculation or prediction of the specific properties (electrical, mechanical, thermal, optical, ...) of the solid by averaging over the many particles contained in the solid. This theoretical scheme is known as statistical mechanics whose solution is the distribution function of the particle number. The distribution function is the probability of finding the particle in a particular range of velocity and direction, between (v_x, v_y, v_z) and ($v_x+dv_x, v_y+dv_y, v_z+dv_z$), and in a corresponding range of kinetic energy between E and E+dE. This number probability distribution function then allows us to calculate or predict the velocity and kinetic energy of the system of particles averaged over the many particles of the system. The averaged velocity and kinetic energy are responsible for the observed properties of the solid and for the response of the solid to an externally applied force, such as a current by an applied voltage, or a displacement or contraction by an applied force or hydrostatic pressure. When the interparticle distance is large or the particle density is low, the identity or the spin of the particle is unimportant in the formulation of the statistical mechanics to find the velocity and energy distribution functions. This is known as the classical statistical mechanics and the distribution function is known as the Boltzmann distribution function. When the interparticle distance is small or the particle density is high, the particle identity, including the spin, must be taken into account to exclude (Pauli's exclusion principle proposed in 1925) two identical particles (including spin) from locating at the same point in the position-spin space. The additional dependence of the probability or wavefunction and energy on the spin is added to the Schrödinger equation for one or many particles. This statistical theory is known as the quantum statistical mechanics. The particle distribution function at the high particle density is known as the Fermi function for particles with odd integer multiplier of 1/2 spin unit such as electron with spin of 1/2, and the Bose function for particles with even integer multiplier of 1/2 spin unit.

Thus, the Schrödinger equation provides the probability of finding a particle at a specified velocity, position and time interval to take into account the uncertainty during measurement by scattering, and the Fermi-Bose distribution functions are the statistical distributions of the particles at high particle densities in solids to calculate and predict the solid's properties averaged over all the particles composing the solid.

P100.2 What distinguish(es) hypotheses from laws? When does a hypothesis become a law? Are the following rules hypotheses or laws and why: Newton, Coulomb, Ampere, Planck, de Broglie, Bohr atom, tunneling, Fermi distribution, Bose distribution, Shockley diode, MOSCV, SNS diode, Bethe diode, Mott-Schottky diode, NMOS, CMOS? (Answer those which you already know. Others will be described in detail in the following chapters. Do this problem again at the end of the semester.)

A law is a quantitative relationship that is deduced from observations in many similar and different experiments. In the unabridged Random House Dictionary of the English language, 2nd edition (Random House, New York, 1987), "in science . . . a statement of a relation or sequence of (experimental) phenomena invariable under the same conditions." A hypothesis is "a proposition, or set of propositions, set forth as an explanation for the occurrence of some specified group of phenomena, either asserted merely as a provisional conjecture to guide investigation or accepted as highly probable in light of established facts." A hypothesis becomes a law when it is shown by experiment to be a statement of an invariable relation.

Newton:	Law, deduced from experiment.
Coulomb:	Law, deduced from experiment.
Ampere:	Law, deduced from experiment.
Planck:	Hypothesis, conjecture supported by experiments.
de Broglie:	Hypothesis, conjecture supported by experiments.
Bohr atom:	Hypothesis, conjecture supported by existing experiments.
tunneling:	Law, deduced from experimental I-V and other measurements.
Fermi distribution:	Hypothesis, conjecture supported by experiments.
Bose distribution:	Hypothesis, conjecture supported by experiments.
Shockley diode:	Law, deduced from experimental I-V and other measurements.
MOSCV:	Law, deduced from experimental C-V measurements.

SNS diode: Law, deduced from experimental I-V measurements.

Bethe diode: Law, deduced from experimental I-V measurements.

Mott-Schottky diode: Hypothesis, conjecture deduced from experiments but proven incorrect by later experiments and a correct theory by Bethe.

NMOS: Circuit containing nMOST. The response of a specific NMOS circuit is a law and reproducible in repeated measurements. The response is predictable by theoretical (or circuit) analysis.

CMOS: Circuit containing both nMOST and pMOST. The response of a specific CMOS circuit (with given transistors, resistors, capacitors, etc.) is a law and reproducible in repeated measurements. The response is predictable by theoretical (or circuit) analysis.

P110.1 What is the most important fundamental parameter that distinguishes solid, liquid and gas and what are its ramifications? (Hint: A length.)

The fundamental parameter that distinguishes solid, liquid, and gas is the correlation length. It defines the dimension of the region in which the material exhibits a regular arrangement or location-ordering of the atoms. A crystalline solid has long-range order, i.e. the location of many atoms in a large volume are on a periodic lattice, and hence the crystalline solid has a correlation length that is large compared with the interatomic separation, which is another way of saying that the atoms' spacings are constant and the atoms are located on a periodic lattice extending over a large if not the entire volume of the crystalline solid. The atomic density in the crystalline or noncrystalline solid is high, meaning the atoms are packed closely together. This tends to prevent the atoms in the solid from changing location and diffusing from one part of the solid to another so the atomic diffusivity is low. In a liquid, the atoms and molecules are far apart and more apt to change location because their interparticle Coulomb force at the large particle separations are too weak to keep the particles in fixed locations. But a few of the particles in the liquid are close to each other and form a nearly unique structure or geometry of ordered atomic arrangement. Thus, liquids have a short-range order and short correlation length which give them the liquidus properties. In gases, the atoms are very far part and so are moving very rapidly because of the weak interatomic or intermolecular binding forces at the large separations. The atoms or molecules change location rapidly from the random collision or scattering between them even though the interatomic and intermolecular separation is large but the close encounter collisions are still rampant. The ramifications of the correlation length are its relation to the rate at which atoms change location or diffuse, and to the attenuation of a disturbance (electrical, mechanical) by scattering and dissipation. For example, the rate of diffusion is low in crystalline solid but high in gas. Attenuation of an electrical signal is low in a crystalline semiconductor and higher in an noncrystalline or amorphous semiconductor due to the increased random scattering of the conduction or current-carrying electrons which randomizes the electron velocity (current is proportional to velocity in the current flowing direction) and the increased rate of dissipation of the kinetic energy of the current-carrying electrons during scattering. These are represented by the electrical resistance of the material.

P111.1 Review the classifications of solid and in what engineering areas are they used?

The classification schemes discussed in section 111 are grouped by: geometrical (crystallinity vs imperfection), purity (pure vs impure), electrical (electrical conductivity), and mechanical (binding force).

Geometrical (crystallinity vs imperfection): chemical engineering, material science and engineering including ceramic and metallurgy engineering, and electrical engineering (solid-state electronics).

Purity (pure vs impure): chemical engineering, electrical engineering (solid-state electronics), material science.

Electrical (electrical conductivity): electrical engineering.

Mechanical (binding force): mechanical engineering, civil engineering, chemical engineering, aeronautical engineering, materials science.

P120.1 Itemize the reasons why crystallinity and semiconductivity are needed to make transistors.

Reasons for crystallinity:

- Non-crystalline semiconductors contain physical defects.
- Physical defects are structural imperfections or displacement of host atoms.
- Physical defects are electron and hole traps.
- Electron and hole disappear when they recombine at traps.
- High trap density gives short electron-hole lifetime.

– Electrical signal carried by electron is attenuated if lifetime is short.
– Electrical attenuation reduces the transistor's current amplification factor.

Reasons for semiconductivity:

– Impurities are added to semiconductors
  * To control the magnitude of the electrical conductivity.
  * To give two conductivity types:
    n-type conduction by electrons.
    p-type conduction by holes.
  * Transistors require two conductivity types
    to form highly nonlinear rectifying p/n junctions.

P131.1 Identify two primitive and two non-primitive cells of the two-dimensional square lattice which are not given in Fig. 131.1.

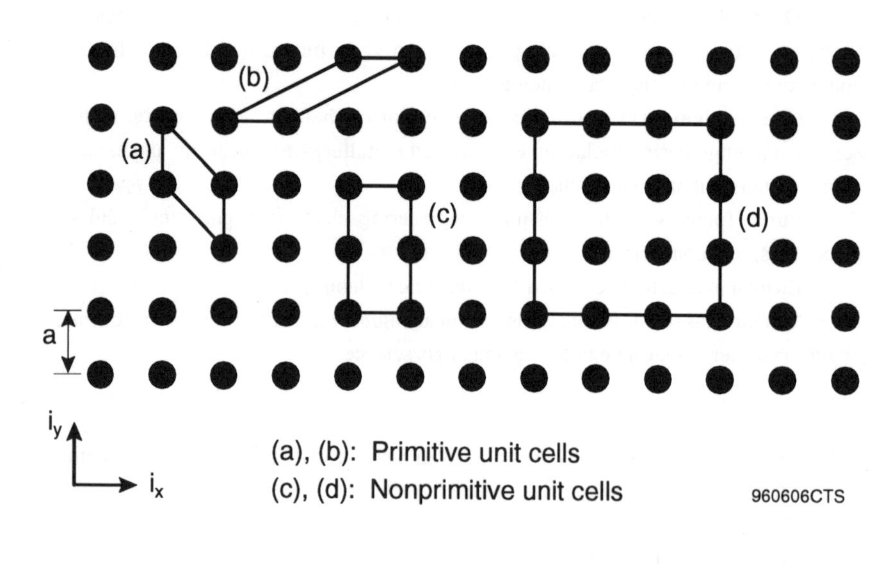

P131.2 Draw the two dimensional lattice whose primitive translation vector of the lattice is $2a\mathbf{i}_x + a\mathbf{i}_y$. Show two primitive and two non-primitive unit cells, one of each should be non-rectangular.

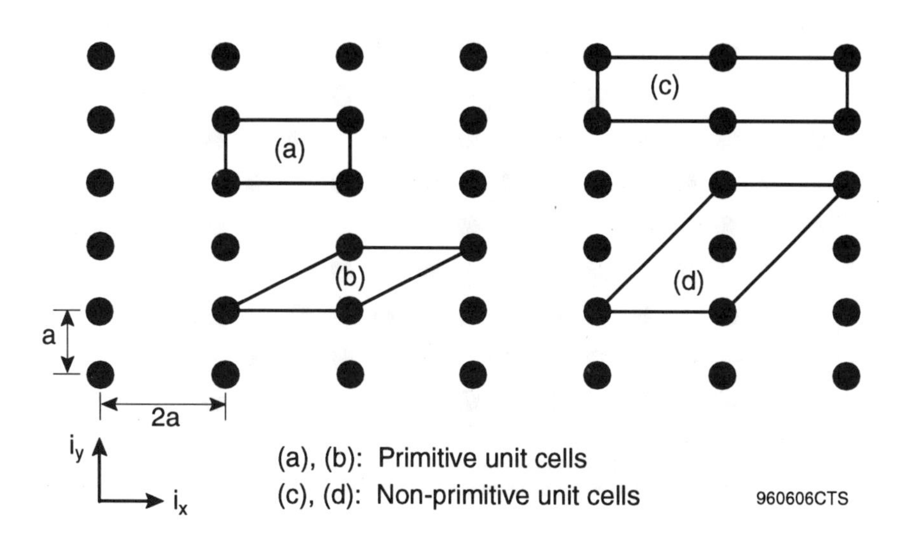

P132.1 Draw the (100), (110) and (111) planes of a face-centered cubic lattice.

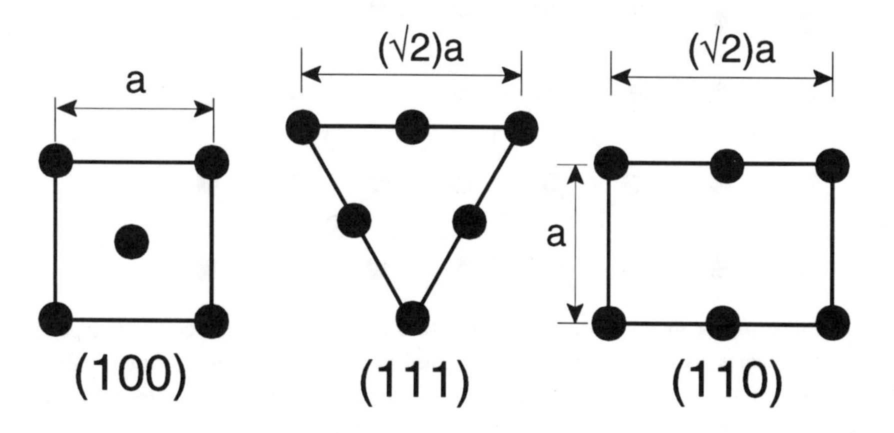

P131.3 Identify two primitive and two non-primitive cells of a two-dimensional face-centered square lattice. Give the primitive unit vector and draw it on the figure. Is this a primitive lattice or a composite lattice of two simple lattices?

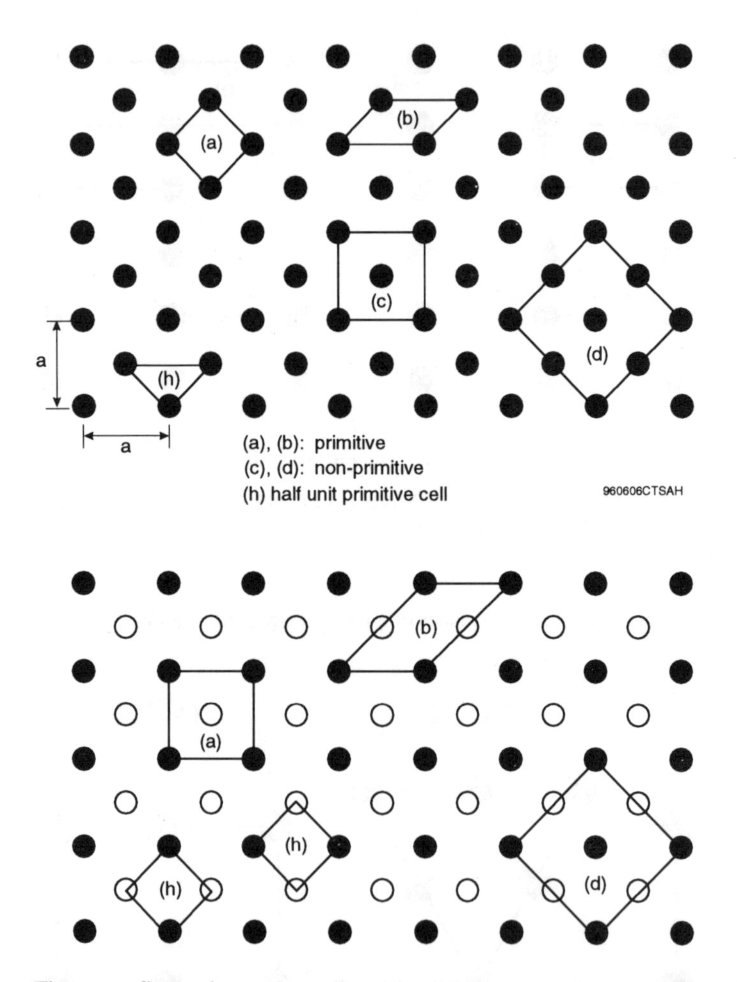

The upper figure shows a two-dimensional face-centered square lattice of one type of atoms which is just a simple square lattice turned 45°. However, if there are two kinds of atoms as shown in the lower figure, then it is a composite lattice of two simple square lattices. The primitive unit cells of the two lattices are different.

P133.1 Obtain the atomic density expression of the two-dimensional rectangular lattice given in P131.2 using four unit cells similar to those selected for the square lattice in Fig. 133.1.

Consider the four unit cells (a), (b), (c), and (d) shown in Figure P131.2. The results are tabulated below. It is evident that the atomic density is independent of the unit cell geometry selected as anticipated.

Unit Cell Type	Cell Area	Number of Atoms per Cell	Atomic Density
(a)	$2a^2$	1	$1/2a^2$
(b)	$2a^2$	1	$1/2a^2$
(c)	$4a^2$	2	$1/2a^2$
(d)	$4a^2$	2	$1/2a^2$

P133.2 Calculate the atomic density of Ga atom, As atom, and GaAs atom-pair in GaAs.

In a zinc blende semiconductor crystal lattice, one type of atom occupies a fcc lattice and the other type of atom occupies the other fcc lattice.

Unit Cell Type	Cell Volume	Number of Atoms per Cell
Ga atom	a^3	8 corner atoms × 1/8 = 1 atom
FCC		6 face-centered atoms × 1/2 = 3 atoms
		Ga atom density = 4 atoms/a^3
As atom	a^3	8 corner atoms × 1/8 = 1 atom
FCC		6 face-centered atoms × 1/2 = 3 atoms
		As atom density = 4 atoms/a^3
GaAs atom pair	a^3	8 corner pairs × 1/8 = 1 pair
FCC		6 face-centered pairs × 1/2 = 3 pairs
		GaAs pair density = 4 pairs/a^3

P133.5 Oxygen causes indirect problems in controlling the electrical properties of silicon transistors and integrated circuits as described in section 134. The segregation coefficient of oxygen in Si is unity. Can oxygen impurity be removed from Si using the zone refining technique? How can oxygen be removed during the growth of a silicon crystal?

Because the segregation coefficient of oxygen in Si is unity, the oxygen concentration is the same in the solid and the melt. Therefore, the zone refining technique cannot be used to reduce the oxygen concentration. However, oxygen can be removed by heating the silicon in an inert gas (He, Ar) ambient or in vacuum (such as in vacuum float-zone crystal growth or zone-refining furnaces), causing outdiffusion of the oxygen in the solid silicon to the solid surface and evaporation into the gas phase or vacuum ambient which is then evacuated. Silicon crystal grown in vacuum ambient has much lower oxygen concentration ($\sim 10^{16}cm^{-3}$) than grown from Si melt in a fused quartz (SiO_2) crucible ($\sim 5\times10^{18}cm^{-3}$).

P140.1 Answer the question posed in Problem 110.1 again but this time, give a more detailed and less qualitative or more quantitative but concise (itemized) description of the fundamental reasons.

[What is the most important fundamental parameter that distinguishes solid, liquid and gas and what are its ramifications? (Hint: A length.]

The fundamental factor discussed in Section 140 that distinguishes solid, liquid, and gas is the particle density.

The consequence of the high particle density in solid is: (1) interparticle distance is very small, about 2 angstroms, (2) the force on a single particle is due to all the other, $10^{23} - 1$, particles, and (3) the rate of collision between the particles is high.

The small interparticle distance cited in (1) means that the instantaneous position of the particle can no longer be defined and the deterministic classical mechanics must be replaced by the probabilistic quantum mechanics.

The interaction between all of the surrounding particles cited in (2) and the high rate of collision cited in (3) indicate that the average motion of many particles is the quantity measured during experiment. Therefore, the statistical properties of a large number of particles are necessary to explain the measured electrical characteristics which is the realm of statistical mechanics.

P141.1 Bohr's first postulate in his hydrogen atom model states that the orbital angular momentum of the bound electron is quantized (or discrete). After (141.4) in the text, a statement is made that this has a very simple geometrical interpretation. Give the graphical demonstration for n=1 and n=4 (easier to draw than n=1) that is anticipated in the text.

The picture on the right depicts a standing sine wave drawn on the circle whose radius is the Bohr radius so that the circumference of the circle is exactly equal to four deBroglie wavelengths. The figure is a sketch of the standing wave whose wavelength is one fourth the circumference of the circle.

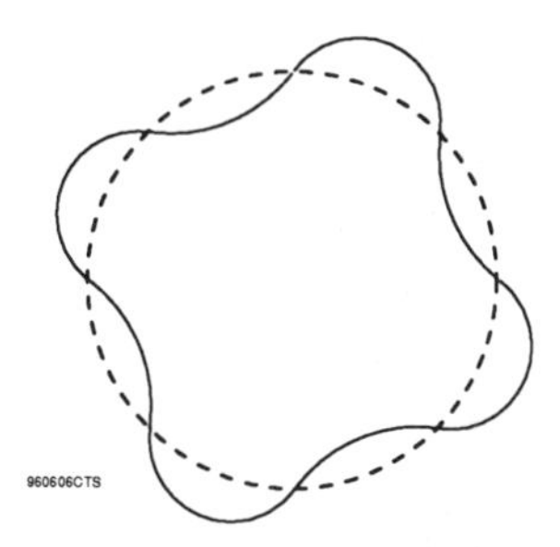

P141.1 Show that Fig.141.5 is drawn correctly with respect to the text statements.

Page 46 states that electric field points in the x-direction of the (x,y,z) coordinates but does not specify positive or negative x-direction. Figure 141.5 shows $E_x < 0$ and a force on the electron in the +x direction.

P141.2 Are the kinetic and potential energies spatially constant for an electron bound at the n-th orbit with energy E_n in Bohr's hydrogen atom model? Are the answers valid in the wave model described by Schrödinger's equation in section 156?

The total energy or the sum of the kinetic and potential energies is a constant. The kinetic and potential energies may not be individually.

P141.4 A ball of 2000 grams is moving at 150 meter/sec. A 3000 pound automobile is moving at 65 mile/hour. An oxygen molecule in our air is moving at 10^7 cm/s. What is the kinetic energy (in joule and electron-volt), de Broglie wavelength in meters, and Planck frequency in Hz of these moving objects? Which particle may require wave-quantum mechanics to explain its motion and why?

Answers for similar problems are on p.13 of FSSE-SG and for this problem below.

Key equations:

$KE = (1/2)mv^2$ classical law

$\lambda = h/p$ de Broglie wavelength

$E = hf$ Planck's energy quantum of a wave

(1) **Ball**

Given: m = 2000g = 2kg, v = 150 m/s

Computed:

$KE = (1/2)(2\text{ kg})(150\text{ m/s})^2 = 2.25\times10^4\text{ J}$

$KE = (2.25\times10^4\text{ J})(1\text{ eV}/1.602\text{x}10^{-19}\text{ J}) = 1.40\times10^{23}\text{ eV}$

$\lambda = (6.6262\times10^{-34}\text{ J}\bullet\text{s})/[(2\text{ kg})(150\text{ m/s})] = 2.21\times10^{-36}\text{ m} = 2.21\times10^{-24}\text{ A}$

$f = (2.25\times10^4\text{ J})/(6.6262\text{x}10^{-34}\text{ J}\bullet\text{s}) = 3.40\kappa10^{37}\text{ Hz}$

(2) **Car**

Given: m = (3000 lb)(1 kg/2.2 lb) = 1364 kg

v = (65 mile/hour)(1.609km/mile)(1000m/1km)(1hr/3600s) = 29.058 m/s

Computed:

$KE = (1/2)(1364\text{ kg})(29.058\text{ m/s})^2 = 5.76\times10^5\text{ J}$

$KE = (5.76\times10^5\text{ J})(1\text{ eV}/1.602\times10^{-19}\text{ J}) = 3.59\times10^{24}\text{ eV}$

$\lambda = (6.6262\times10^{-34}\text{ J}\bullet\text{s})/[(1364\text{kg})(29.058\text{m/s})] = 1.67\times10^{-38}\text{ m} = 1.67\times10^{-28}\text{ A}.$

$f = (5.76\times10^5\text{ J})/(6.6262\times10^{-34}\text{ J}\bullet\text{s}) = 8.69\times10^{38}\text{ Hz}$

(3) **O_2 molecule**

Given: $m_{O2} = 2(16\text{amu})(1.66\times10^{-24}\text{g/amu})(1\text{kg}/1000\text{g}) = 5.3126\times10^{-26}\text{ kg}$

$v = (10^7\text{cm/s})(1\text{m}/100\text{cm}) = 10^5\text{ m/s}$

Computed:

$KE = (1/2)(5.3126\times10^{-26}\text{ kg})(10^5\text{ m/s})^2 = 2.66\times10^{-16}\text{ J}$

$KE = (2.6563\times10^{-16}\text{ J})(1\text{ eV}/1.602\times10^{-19}\text{ J}) = 1.66\times10^3\text{ eV}$

$\lambda = (6.6262\times10^{-34}\text{ J}\bullet\text{s})/[(5.3126\times10^{-26}\text{ kg})(10^5\text{ m/s})]$

$= 1.25\times10^{-13}\text{ m} = 1.25\times10^{-3}\text{ A}.$

$f = (2.66\times10^{-16}\text{ J})/(6.6262\times10^{-34}\text{ J}\bullet\text{s}) = 4.01\times10^{17}\text{ Hz}$

(4) **O_2 molecule**

Given: $m_{O2} = 2(16\text{amu})(1.66\times10^{-24}\text{g/amu})(1\text{kg}/1000\text{g}) = 5.3\times10^{-26}\text{kg}$

$v = (10^4\text{cm/s})(1\text{m}/100\text{cm}) = 10^2\text{ m/s}$

Computed:

$KE = (1/2)(5.3126\times10^{-26}\text{ kg})(10^2\text{ m/s})^2 = 2.66\times10^{-22}\text{ J}$

$KE = (2.6563\times10^{-22}\text{ J})(1\text{ eV}/1.602\times10^{-19}\text{ J}) = 1.66\times10^{-3}\text{ eV}$

$\lambda = (6.6262\times10^{-34}\text{ J}\cdot\text{s})/[(5.3126\times10^{-26}\text{ kg})(10^2\text{ m/s})]$

$= 1.25\times10^{-10}\text{ m} = 1.25\text{A}$

$f = (2.66\times10^{-22}\text{ J})/(6.6262\times10^{-34}\text{ J}\cdot\text{s}) = 4.01\times10^{11}\text{ Hz}$

The motion of the ball and car are adequately described by classical Newtonian mechanics since their body are very massive, their kinetic energy very large, and their deBroglie wavelength very small, compared with those of a measuring photon or particle which scatters off the massive body during measuring. The uncertainty in position determination during observation via scattering is of the order of λ from the uncertainty principle, $\Delta x\Delta p \sim h$, or $\Delta x \sim h/\Delta p \sim \lambda$, and hence negligible. The motion of the O_2 molecule with a high velocity in (3) can also be determined by Newtonian mechanics for the same reasons, however, the 1000-times slower moving O_2 molecule in (4) would require Schrödinger equation to give the probability of finding the O_2 for exactly the opposite reasons, large λ and small K.E.

P141.8 Draw the potential energy curve of an electron in the field of two motionless protons which are held at x=-a and x=+a. Use r=∞ as the reference potential energy. Label also the kinetic energy of the electron if its total energy is positive, E>0.

The reference potential, $V(\infty) = V(-\infty) = 0$, is chosen since $1/r \to 0$ as $r \to \pm\infty$. The electron potential energy due to the proton at x=−a and x=+a are $V_{proton1}(x) = -q^2/(4\pi\varepsilon_0|x-(-a)|)$ and $V_{proton2}(x) = -q^2/(4\pi\varepsilon_0|x-a|)$. The total electron potential energy is the sum:

$$V(x) = V_{proton1}(x) + V_{proton2}(x)$$
$$= -q^2/(4\pi\varepsilon_0|x+a|) - q^2/(4\pi\varepsilon_0|x-a|)$$
$$= -(q^2/4\pi\varepsilon_0)\times(|x+a|^{-1} + |x-a|^{-1})$$

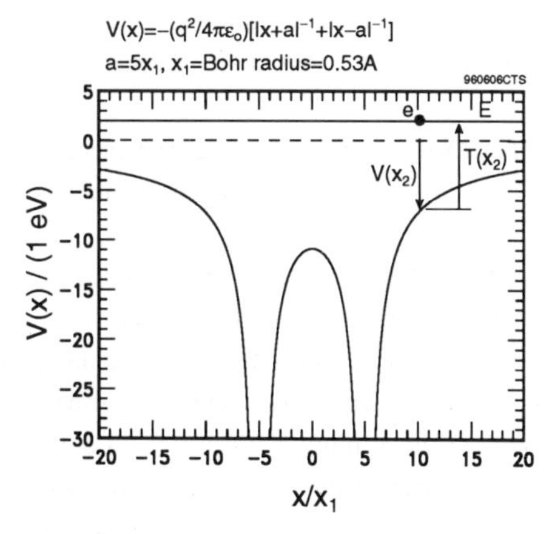

P141.9 In the two-proton problem above, do you expect the ground state energy of the electron to be more negative (more tightly bound) or more positive than the electron in the one-proton or hydrogen ground state? Why? Label the kinetic energy of the electron bound to the two-proton ground state.

More tightly bond because the two protons exert twice as much attractive force on the negatively charged electron, if the two protons are located at the same point in space or a=0. The separation of the two protons will reduce the attractive force from twice as much but it is still more than just the attractive force from one proton.

P141.10 The two lowest-energy wavefunctions of the electron bound to the two protons are symmetric and antisymmetric. Sketch and explain why one has a lower energy while the other has a higher energy.

The symmetric wavefunction peaks at the origin, x=0, to satisfy the space symmetry condition, $\psi_S(x) = +\psi_S(-x)$. The antisymmetric wavefunction must have null amplitude at the origin to satisfy the antisymmetry condition, $\psi_A(x)=-\psi_A(-x)$, thus, $\psi_A(x=0)=0$ and it peaks away from the origin. The potential energy at and around x=0 is large and negative where the symmetric electron spends more time. Thus, the total energy of the symmetric bound state wavefunction is larger (or more negative) than that of the antisymmetric wavefunction. The $\psi(x)$ vs x are similar to the lower two in Fig.155.1(b). A similar result for a more complex problem of many electrons in a crystal is given in Fig.181.3 and P181.8 for the same physical reason, i.e. the electrons concentrated at the atomic core are more tightly bond and hence with large and more negative total energy corresponding to those in the valence band.

P141.13 Can a true bound electron energy state exist in the potential energy given in Fig. 141.5? Why? (Hint: No, tunneling.)

True bound electron energy states cannot exist in Fig.141.5 which is the potential energy diagram of an electron around a proton in the presence of an applied electric field because if the bound state exists, the bound electron will eventually tunnel out of the potential well and hence cannot be permanently bound, like an electron around a proton without the applied electric field, i.e. in hydrogen atom. Mathematically, a bound solution of the Schrödinger equation does not exist if the potential energy term consists of two terms, $Fx - A/r$, where Fx comes from the applied force or electric field along the x-direction. In practice, the electric field is small, and the classical-forbidden potential barrier is thick and the tunneling rate is so small that during any practical observation time, the electron is attached to the positive charge which gives the $-A/r$ Coulombic attractive potential. Since the Bohr ground state energy is 13.6eV and the Bohr radius is 0.52A, the electric field due to the proton is about $13.6/0.52\times10^{-8} \sim 10^{9}$V/cm. So, the applied electric field has to be extremely high to cause rapid tunneling of the electron bound to the proton at the ground state of the hydrogen atom. Figure 141.5 shows to scale an electric field of $10V/20r_1 \sim 10^8$V/cm which is about the practical upper limit that can be produced (100 Mega-Volt per centimeter) by the very reason that gas would ionize and the electrode atoms would be ripped off to prevent sustaining or even producing such a high electric field in the laboratory. The forbidden barrier thickness at -13.6eV estimated from extending the figure in the +x direction is $\sim 13.6/10^8$ or $a \simeq 13.6$A and the barrier height is $E_{Barrier} = E_{peak} - (-13.6) \simeq (-7) + 13.6 \simeq 6.6$eV. This is a very thick and tall barrier, and the tunneling probability calculated from (153.11) assuming a square barrier, $\exp[-2(a/\hbar)\sqrt{2mE_B}]$ or (154.1A) assuming a triangular barrier is extremely small. The students may carry this estimate out by plugging in the numbers to determine the tunneling transition probability.

P142.1 Why do we pick the negative sign for **p** instead of positive sign?

In order that a traveling wave with positive velocity is traveling in the positive x-direction. See graphical solutions for the P142.5.

P142.4 Construct a traveling wave moving in the negative x direction.

$$y(x,t) = Y(0)\cos(\omega t + kx) \quad \text{(P142.4A)}$$

The velocity of this traveling wave is negative, i.e. the point of constant phase is moving in the negative direction with time as demonstrated by

$$\text{Phase} = \Theta = (\omega t + kx) = \text{constant}.$$

Thus, phase velocity is $dx/dt = - d\omega/dt < 0$.

One can also graphically illustrate this by drawing the cosine wave versus x at t=0 and a later time $t=t_1$ which is demonstrated in the next problem for a travelling wave moving in the positive x direction.

Other examples are the following wavepackets that could represent a group of electrons moving in the negative direction or positive direction (change the + sign to − sign below). The first does not decay while the second decays and spreads out.

$$y(x,t) = Y(0)\exp[-(x+vt)^2/4Dt_0] \quad \text{(P142.4B)}$$

$$y(x,t) = Y(0)(t+t_0)^{-3/2}\exp[-(x+vt)^2/4Dt] \quad \text{(P142.4B)}$$

P142.5 Which of the following are traveling waves and standing waves? Use sketch at t=0 and $t=t_1$ to illustrate your answers. (a) $\exp[i(kx-\omega t)]$; (b) $\cos(kx-\omega t)$; (c) $\cos(kx)\cos(\omega t)$; (d) $\exp[-(\alpha x-\omega t)]$; (e) $\exp(-\alpha x)\cos(\omega t)$; (f) $\exp[-\alpha(x-x_0)^2-i\omega t]$. Let $k>0$, $\omega>0$ and $\alpha>0$.

Traveling waves: (a), (b), (d)
Standing waves: (c), (e), (f)

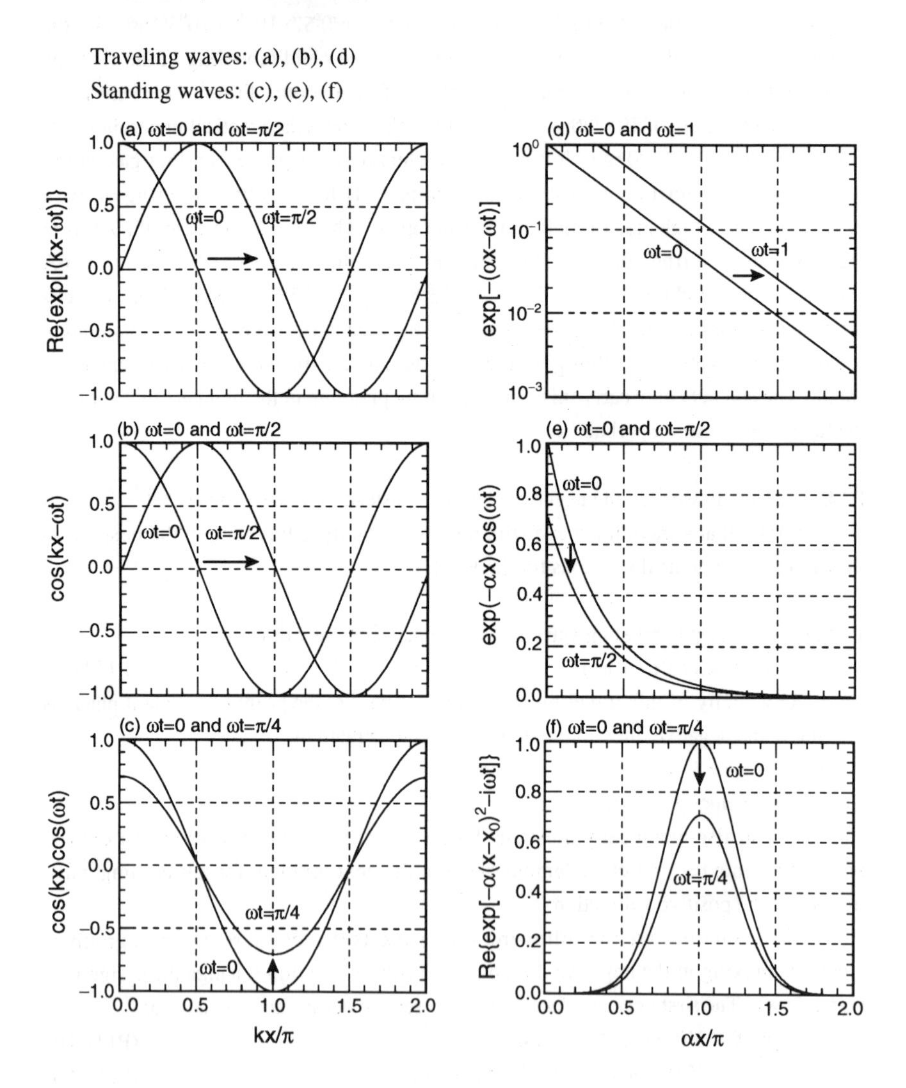

P161.2 The electron affinity to an isolated neutral Si atom is 1.39eV while to a neutral Si crystal it is 4.02eV. Why is it larger in Si crystal? Give a simple explanation based on electrostatic or Coulomb force.

The electron affinity is the energy required to remove an electron from the subject. It is smaller in an isolated neutral Si atom because the to-be-removed electron is bound to the Si^{+4} atomic core which is screened by the other three nearby valence electrons so the net positive charge is much less than +4 and the attractive force is hence much less than that due to a +4q charge. In addition, the electron-electron repulsion energy between this and other three valence electrons reduces the binding. In the Si crystal the electron affinity is larger, almost by a factor of three, because all the four valence electrons of each Si atom are spread out over the entire crystal owing to the small interatomic distance or the presence and the proximity of the neighboring Si^{+4} ions so that the +4q charges and their Coulomb attraction force on the to-be-removed electron is not as much reduced as in an isolated neutral Si atom.

P172.1 Why are the valence electrons the most important on influencing the electrical properties of silicon? Why are the core electrons not as important? Why are the core electrons important to distinguish Si, Ge, GaAs and others?

The valence electron radii are comparable to the interatomic spacing and hence are spread out over the entire crystal. Thus, they are more responsive to the applied electric field to conduct electricity. The core electrons are tightly bound to each atomic core and hence are not responsive to the applied forces. Thus the electrical properties are mainly determined by the responses of the valence electrons of the atoms composing the solid to the applied forces.

Although the core electrons are not directly responsible for electrical conduction, their spatial distribution around the positively charged nucleus determine the net positive charge seen by the valence electrons. This core-electron-screened attractive force from the protons in the nucleus influence the mobility of the valence electrons moving in an electric field which is represented by an effective mass. The nuclear charge and the spatial distribution of its core electrons are different for Ga, As and Si cores and this difference is responsible for the different electrical and electro-optical properties exhibited by these semiconductors.

P172.4 If all the valence electrons in silicon are in the valence band, what is the minimum photon energy that is just enough to release an electron from silicon into vacuum?

From Figure 172.2, this is

$$h\nu_{minimum} = \text{electron affinity} + \text{energy gap}$$
$$= \chi_{Si} + E_G$$
$$= 4.02\text{ eV} + 1.18\text{ eV} = 5.2\text{ eV}$$

P172.8 Draw an energy band diagram, especially near the semiconductor-vacuum interface or the semiconductor surface, in the presence of an electric field.

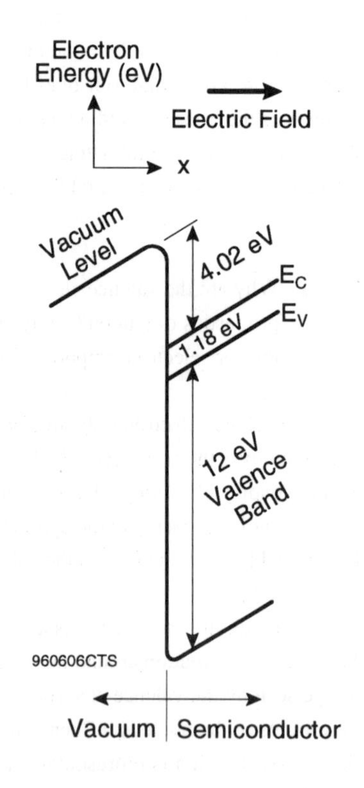

The effect of an electric field on the energy band diagram may be obtained from the relations between electric field, electric potential and electron energy: Electric field(V/cm) = –dV/dx;
Electron energy(eV) = –qV; and
V = [Electron energy(eV)]/(-q).
These give:
Electric field(V/cm) =
= –(d/dx)(Electron energy/(–q)
= (1/q)(d/dx)(Electron energy).
Thus, an applied electric field tilts the energy band by a slope given above.

Chapter 2
HOMOGENEOUS SEMICONDUCTOR AT EQUILIBRIUM

OBJECTIVES

Understand two basic concepts

- Homogeneity
- Equilibrium
  - Electronic Equilibrium
  - Atomical or Atomicitic equilibrium
    * Atomic Equilibrium (Chemical Equilibrium)
    * Ionic Equilibrium (Electrochemical Equilibrium)
    * Thermal equilibrium (atomic vibrations, heat transfer)
    * Mechanical equilibrium (many-atom massive body equilibrium)

Know and Use Semiconductor Parameters

- Intrinsic Carrier Concentration, n_i
- Donors and Acceptors
- Charge States of Donors and Acceptors
- Trapped Electrons and Holes and Binding Energies
- Fermi Distribution of Electron Kinetic Energy
- Boltzmann Distribution of Electron Kinetic Energy
- Density of Electron States or Energy Levels
- Intrinsic and Extrinsic Fermi Energy Levels
- Extrinsic Electron and Hole Concentrations

P202.1 Describe the partial equilibrium conditions of a piece of Si sitting on top of a desk in a room at a space-time constant temperature of 300K.

As explained in Section 202, equilibrium means that the macroscopic averaged properties of a body made up of particles under observation do not change with time. The piece of Si on top of the desk is not in atomic or chemical equilibrium because of the concentration gradient of silicon atoms and desk atoms at the interface. However, the interdiffusion of the silicon and desk atoms through the surface contact of the two solid bodies at room temperature is very small during the observation period, therefore, the two bodies are in approximate atomic or chemical equilibrium. If there are impurity ions in the silicon and not in the desk, the two contacting bodies are not in ionic equilibrium due to the concentration gradient of the impurity ions at the contact interface. However, the ionic current is negligible during the measurement period because of the very small diffusion rate at room temperature, thus, the piece of silicon and the desk are in approximate ionic equilibrium. Assuming that the silicon and desk have been in contact for a sufficiently long time such that the average kinetic energy of the silicon atoms and desk's ions are equalized or the same, then it is in thermal or thermodynamic equilibrium which is conventionally measured by their temperatures (identical).* Because the silicon is motionless sitting on top of the desk with negligible atomic and ionic interdiffusions of the particles composing the two solid bodies during the period of measurement, they are in mechanical equilibrium. (*Students may get an exercise by giving the solution for the problem of a thermocouple measuring the temperatures.)

P202.2 A piece of pure silicon is immersed in pure boiling water. What are the essentially equilibrium parameters and the dominant nonequilibrium parameters?

If the average kinetic energy of the silicon lattice measured by a temperature, T_1, is different from the average kinetic energy of the boiling water measured by a temperature, T_2, the dominant nonequilibrium parameter when the piece of silicon is immersed into the boiling water is the thermal nonequilibrium. In addition, due to the spatially different concentration of the silicon atoms and the water molecules, there is an atomic or chemical and ionic or electrochemical nonequilibrium via diffusion and drift, and reaction. The water molecules or its ionized parts, $H_2O^{\otimes} \rightarrow H^{\oplus} + OH^{\ominus}$,may diffuse into silicon due to the concentration gradient and drift in the built-in electric field. They may then react with the interior Si atoms or stay at interstitial locations in the silicon lattice. The silicon atoms on the silicon surface may also react with the water molecules ($Si+2H_2O \rightarrow SiO_2 + 2H_2$), so there is a continuous change of the silicon body size whose surface is covered by a growing layer of silicon dioxide of increasing thickness. If the silicon is motionless in the boiling water, it is in mechanical equilibrium with the water, except the small weight and size change from surface oxidation. However, the water molecules are evaporating via boiling and hence is not in an atomic equilibrium state.

P202.3 A solar cell panel is attached to the exterior of an orbiting satellite which can be considered a vacuum. The panel is initially in the shadow. It is suddenly exposed to the sunlight. Describe the transient events that lead to thermal equilibrium of the atoms and the remaining nonequilibrium events. Disregard the electrons and holes for this problem. Consider also the electronic equilibria after optoelectronic devices are studied.

When the cold panel in the shadow is suddenly exposed to the sunlight, the incident photons transfer kinetic energy to the surface atoms, causing the vibration amplitude of the host and impurity atoms of the solar cells to increase about their respective lattice sites. The surface atoms are then no longer in thermal equilibrium with the interior atoms because the surface atoms now have a higher vibrational kinetic energy. The vibrational energy propagates from the surface to the bulk via the collision and scattering of the higher kinetic energy surface atoms with the lower kinetic energy interior atoms inside the solar cell. This is known as thermal conduction or heat transfer. Some energy is also radiated from the surface atoms via emission of photons at infrared energies into the vacuum of space during absorption pf the incident photons from the sun light. These energy transfer events could be measured by a thermocouple which extract a small percentage of energy by thermal conduction just described through the contact interface between the thermocouple and the solar cell panel to drive a mechanical (meter) or color indicator that is calibrated to give temperature readings. The solar cell panel temperature reaches constant value, higher than that in the shade, when the steady-state is attained, that is, the rate of energy change of the three processes (thermal conduction, radiation, and absorption) exactly balance so the net rate of energy change is zero.

P202.5 What is the deBroglie wavelength of an electron at the average kinetic energy k_BT at T=300K? [Why not $(3/2)k_BT$?] This is known as the **thermal de Broglie wavelength** of an electron. How many valence electrons are there in a thermal de Broglie cube of Si whose edge is the thermal de Broglie wavelength? What is the fluctuation in electron number in such a cube? What is the significance of localization of an electron by a wave packet in a thermal de Broglie cube?

The thermal de Broglie wavelength at energy k_BT can be computed from $\lambda=h/p$, $p=mv$, and $k_BT=(1/2)mv^2 = p^2/2m$ which gives $p=\sqrt{(2mk_BT)}$. Thus, $\lambda=h/\sqrt{(2mk_BT)}$.

We can either use the cgs, MKS or the energy unit for the constants:

$h = 6.62618\times10^{-27}$ erg-sec $= 6.62618\times10^{-34}$ J-s or

$h/q = 4.13570\times10^{-15}$ V-sec

$k_B = 1.38066\times10^{-16}$ erg/K $= 1.38066\times10^{-23}$ J/K or

$k_B/q = 8.61734\times10^{-5}$ eV/K

$m = 9.10953\times10^{-28}$ g $= 9.10953\times10^{-31}$ kg

$m/q = 5.58568\times10^{-16}$ V/(cm/s)^2

This gives a thermal deBroglie wavelength at 300K of $\lambda = 76.95711\times10^{-8}$ cm or ~ 77A. The silicon valence electron density computed from the cubic unit cell containing 8 Si atoms or 32 valence electrons is $32/a^3 = 32/(5.43\times10^{-8}\text{cm})^3 = 1.998\times10^{23}\text{cm}^{-3}$. Thus, in a de Broglie cube of 77^3A^3 volume, the number of valence electrons $2\times10^{23}\text{cm}^{-3}\times77^3\times10^{-24} = 9.1\times10^4 \approx 10^5$ or about one hundred thousand electrons. The number fluctuation is about $\sqrt{10^5} \sim 300$ electrons.

The average valence-electron spacing $(2\times10^{23})^{-1/3} \simeq 1.71$A is much smaller than the thermal de Broglie wavelength of an electron with energy k_BT which is a measure of the position uncertainty or fluctuation of an electron excited to a kinetic energy of k_BT. Thus, this thermal electron cannot be localized at the scale measured by the interatomic spacing of silicon (2.35A) or the inter-valence electron spacing (1.71A) in silicon.

P202.6 A piece of crystalline silicon slice contains an impurity whose concentration is given by $N_{Impurity} = ax$ where x is alone the direction perpendicular to the surfaces of the Si slice and 'a' is a constant. The Si is held motionless and immersed in a stationary ambient gas. The two surfaces of the Si slice are in perfect thermal contact with the ambient gas which is held at 300K and which can be considered as an infinite heat sink and heat source. Is the Si slice at thermodynamic equilibrium? Does it approach several partial equilibrium conditions? Give reasons.

The impurity concentration varies with position, therefore the slice is not homogeneous and is strictly not at equilibrium because the impurity atoms will migrate from the higher concentration region to the lower concentration region by diffusion due to the concentration gradient. However, at room temperatures, the diffusion rate of the impurity is so small that during the practical observation time, the number of atoms diffused is negligible (Chapter 3 will give quantitative estimates to show that impurity diffusion is negligible at 300K but important at high temperatures, > ~1000K.) Thus, the Si slice is at atomic or chemical equilibrium approximately because of the negligible diffusion of the impurity atoms. The silicon slice is also at thermal equilibrium with its surrounding from the conditions described in the problem statement. Since no electric current is specified, the Si slice is also approximately at electronic equilibrium (due to the electrons and holes from the impurity atoms) and ionic equilibrium (due to the ionized impurity atoms which have released their electrons if a donor and accepted valence bond electrons if an acceptor).

Finally, since the Si slice is at thermal equilibrium and it must be electrically neutral as suggested by the conditions described in the problem which cannot allow a charge to be built up, its electron and hole concentrations would vary with position because of the spatially varying impurity concentration in order to preserve electrical neutrality in most part of the Si slice except x=0, i.e. macroscopic space charge density, given by $\rho(x) = q[P - N + N(x)] = q[P - N + ax]$ vanishes. (Here, we assume that the impurity is a donor and all ionized so positively charged.) In addition, the thermal equilibrium condition requires $PN = n_i^2$ = spatially constant. Thus, the electron concentration vary approximately linearly with distance when $x >> 0$ and the hole concentration varies approximately linearly with distance when $x << 0$. This concentration gradient would cause the electron (or hole) to diffuse from the higher concentration region ($x>>0$) to the lower concentration region ($x<<0$) to produce a diffusion current. But the diffused electrons would then charge up the lower concentration region ($x<<0$) negatively and the departed electrons would make the ($x>>0$) region positively charged. The excess positive and negative charges on two sides would produce an electric field that cause the electrons to drift from the $x<<0$ region to the $x>>0$ region. This electron drift current exactly balances the electron diffusion current flowing from $x>>0$ to $x<<0$ region to make the net electron current zero. A similar description can be made for the exact cancellation of the hole drift and diffusion currents. Thus, the inhomogeneous Si slice is at electronic equilibrium also when there is no current flowing into or out of the Si slice from two leads connected to it even though there are four internal currents, drift and diffusion of electrons and holes. (Chapter 3 will describe the underlying physics and the characterization mathematics of the drift and diffusion currents.)

Thus, the Si slice is approximately at partial electronic, chemical (atomic), and ionic (electrochemical) equilibria, and at thermal and mechanical equilibria as specified by the problem statements.

P210.1 How many intrinsic electrons and holes are there in a thermal de Broglie cube at 300K in Si corresponding to an energy equal to k_BT? What is the thermal fluctuation in carrier density in this cube? Can the intrinsic electrons be localized?

The thermal deBroglie wavelength at energy k_BT can be computed from $\lambda=h/p$, $p=mv$, and $k_BT=(1/2)mv^2 = p^2/2m$ which gives $p=\sqrt{(2mk_BT)}$. Thus, $\lambda=h/\sqrt{(2mk_BT)}$. We can either use the cgs, MKS or the energy unit for the constants:

$h = 6.62618\times10^{-27}$ erg-sec $= 6.62618\times10^{-34}$ J-s or
$h/q = 4.13570\times10^{-15}$ V-sec
$k_B = 1.38066\times10^{-16}$ erg/K $= 1.38066\times10^{-23}$ J/K or
$k_B/q = 8.61734\times10^{-5}$ eV/K
$m = 9.10953\times10^{-28}$ g $= 9.10953\times10^{-31}$ kg
$m/q = 5.58568\times10^{-16}$ V/(cm/s)^2

This gives a thermal deBroglie wavelength at 300K of $\lambda = 76.95711\times10^{-8}$ cm or ~ 77A. Take the intrinsic carrier concentration in silicon at 300K as 10^{10}cm^{-3} for a rounded figure, a deBroglie cube of 77^3A^3 volume, then the number of intrinsic electrons in this cube is $10^{10}\text{cm}^{-3}\times77^3\times10^{-24} = 4.56\times10^{-9}$ electrons. Obviously this number is meaningless and the electron cannot be localized. See P202.5 for further discussions.

P223.4 Give the basic physics reason of at least three errors in each of the two bond diagrams of shallow acceptor and donor in semiconductors show below. (See p.223 of FSSE for the diagram.) These errors are frequently found in freshman chemistry, sophomore physics, and junior semiconductor device textbooks as well as reference books. Mixed symbols are used: Shockley (two bars), Lewis (dots), and ours (circle for hole): they are not to be counted as errors.

Many errors (at least four) can be immediately spotted by comparison of the correct bond models given in Figs.222.1 and 222.2 on p.164 of FSSE with the two erroneous bond pictures for the B and P impurity cited in the problem (p.223 of FSSE).
These erroneous pictures were copied from several most popular freshman-sophomore chemistry-physics and junior transistor-device textbooks, and even some advanced device books and semiconductor physics books. The seriousness serves to focus the attention of the students to understand the underlying physics in order to be able to learn device physics well. These are listed below.

(1) The free hole is next to the boron acceptor impurity. This is erroneous. It must be many lattice constants (a=5.43072A for the Si cubic unit cell) away from the boron acceptor in order for it to be a 'free' hole or a hole moving in the valence bond. It must be at least one and better more than twice the bound hole orbit radius (bound radius is about 6×2.35A or 14A) away from the boron acceptor or it will be a bound rather than 'free' hole. Similarly, the same error is repeated for the 'free' electron in the right-side picture.

(2) The word 'free' is obvious inappropriate and fundamentally flawed since 'free' is meant free space while these electrons and holes are in a solid composed of many other electrons and atomic cores.

(3) The two pictures in the problem description on p.223 of FSSE are incorrect even for a hole bound to the negatively charged acceptor impurity ion core or a electron bound to the positively charged donor impurity ion core. They cannot be located so close to the ions because of dielectric screening by the valence electrons which gives a dielectric constant of 11.9 for silicon. (11.8 was used in some examples such as p.339. 11.9 is preferred.) On p.166, the radius of the ground state of the bound electron or hole is estimated and is about 14A which is 14A/2.35A = 6 times the silicon–silicon bond length or Si-Si nearest neighbor distance shown in the two figures of the problem on p.223.

(4) The two figures in the problem on p.223 use the single charge symbol ⊕ for the silicon atom which is wrong. It should be +4 for the Si core as indicated in Figs.222.1 and 222.2 because the four silicon valence electrons in a silicon crystal are no longer localized at the silicon core and they are represented by the four covalent or two-electron bonds using Shockley's notation of one bar for one electron.

(5) The black dot (smaller for boron and larger for phosphorus) are erroneous since they do not specify the sign and magnitude of the electrical charge of the impurity cores which is +3 for boron and +5 for phosphorus as indicated by Figs.222.1 and 222.3 on p.164 of FSSE.

The original reference of these textbook errors was Shockley's transistor classic [299.1]. Shockley had indeed given all the correct pictures at atomic scale of 2.35A inter-silicon distance [as evident in Figs. 1.2, 1.4, 1.5, 1.6] while describing the bond picture for the current carrying electrons and holes in pure Si. He also gave the correct pictures (also at the microscopic scale of 2.35A) for the bound electron and bound hole in Figs. 1.7 and 1.8 attached to the As and B impurity atom respectively, including providing an estimate of bound orbit radius in section 9.2 on p.223-225 and giving a vivid picture illustrating the bound state wavefunction in Fig.1.14. Furthermore, he even used, correctly with deep physics implication, the single charge symbol, ⊕ and ⊖, to represent the 4-valence-electron-shielded +5 donor and +3 acceptor ions when viewed by an electron or a hole at the macroscopic device dimension of yesterday's millimeter-size single transistor or today's submicron (a few 1000A >> Bohr radius) size VLSI transistors in energy band picture (E-x diagram) in Fig.9.2 on p.226 [299.1]. Although Shockley did not combine all the details in one figure such as Figs.222.1 and 222.2 in FSSE, such as the Si^{+4}, As^{+5} and B^{+3} cores in his Fig.1.14 on p.23, all the basic ideas and fundamental physics were illustrated and qualitatively calculated. Unfortunately the artistic translation attempts of recent introductory textbook authors to integrate these basic ideas into a simple picture have run afoul of many crucial parts of the basic physics that mislead the beginners.

P231.1 Why is it that we can write the Boltzmann factor in terms of the total energy, E, such as $\exp(-\beta E)$, while the basis given in the text discussion of this chapter suggests that it should be written in terms of the kinetic energy (KE), $m^{*2}v^2/2$ or $\hbar^2\mathbf{k}^2/2m^*$, such as $(-\beta m^* v^2/2)$ or $\exp(-\beta\hbar^2\mathbf{k}^2/2m^*)$?

This is the result of the spherical energy band model. If the energy band is not spherical, then we must use the full effective mass relationship for the E(k) function, i.e. $E(k) = (\hbar^2/2)[(k_x/m_x)^2 + (k_y/m_y)^2 + (k_z/m_z)^2]$, in order to be able to do the average of the particle distribution over the anisotropic E-k space.

P231.2 Why is the kinetic energy of electron measured from E_C and holes from E_V rather than vacuum level? **Hint**: Apply similar reasoning for the binding energy E_1 near the end of section 223.

As explained in section 223, the band edge E_C (or E_V) is the reference energy to measure the kinetic energy and binding energy of an electron bound to a positively charged donor impurity ion (or a hole bound to a negatively charged acceptor impurity ion), instead of the vacuum level, because the periodic potential energy of the electrons from the Coulomb attractive force of the positively (or negatively) charged host ions is already contained in the effective mass of the electron (or hole) through a mathematical transformation (i.e. Fourier series analysis as indicated in section 181) which gives the new reference for the potential and kinetic energies of the transformed particles or quasi-particles which we call electrons (or holes) for expedient discussion. And these new references are the E_C for electrons measuring upwards (or E_V for holes measuring downwards) in the one-electron (or one-hole) or the one-quasi-electron (or one-quasi-particle hole) energy band diagram. The energy reference of the two quasi-particle species (electron and hole), E_C and E_V, are separated by an energy gap or forbidden energy range. (The anti-particles, such as proton-antiproton, electron-positron, etc. are the exact analogs in vacuum, whose energy gap is much larger and given by the Einstein energy, mc^2. The difference is the vacuum void and the solid with many background host atoms on a lattice.) So, when a silicon atom is replaced by a donor impurity atom, the potential change from the nuclear charge and electron configuration difference between the silicon atom and its donor replacement, is a positive and spatially distributed (roughly the size of the Si or donor atomic core) single excess positive charge which will bind the negatively charged quasi-electron with a binding energy (negative in the solution of the Schrödinger equation) measured from the reference energy for the quasi-electron, E_C, (or quasi-particle hole, E_V, in case of an acceptor impurity). Although the answer to this problem can be perhaps more simply illustrated with a few lines of elementary mathematics (Fourier series analysis as shown in section 181), the explanation given here provides all the basic physics involved via the fundamental electrostatic or Coulomb force.

P231.4 Give the reasons why the following commonly and carelessly made statement is erroneous. "The principle of detailed balance states that in thermodynamic equilibrium every process and its inverse process proceed at equal rate." While this statement is fundamentally deficient, the equality of (231.1) and (231.2) asserted and explained in section 231 is correctly stated and fundamentally flawless.

The two common misconceptions or at least imprecise if not careless description are stated in section 231 of FSSE and specifically on p.171: (i) the principle of detailed balance applies only to a statistically meaningful sample of large enough number of particles whose number is stationary, and (ii) the transition is macroscopic and involves a large number of particles and not individual particle transition. For example, it is not $T_1=T_2$ in (231.1) and (231.2) but their product with the initial and final allowed states $T_1f_1(1-f_2)=T_2f_2(1-f_1)$. Furthermore, T_1 and T_2 are macroscopic averages over the effecting particles that cause these transitions, in this case, the lattice vibrations or phonons. Thus, T_1 and T_2 are not the single particle transition rates as frequently assumed by veteran teachers and beginner students.

P231.5 Which of the following statements are fundamentally correct and which ones are fundamentally incorrect if taken at their face value without implied interpretation? Give the physics reasons. **(a)** Half of the energy levels at E_F is occupied. **(b)** The energy level at E_F is occupied by one electron. **(c)** The probability that the energy level at E_F is occupied is 50%.

(a) Correct. As explained in Section 231 on page 172, at a given temperature T at thermal equilibrium, the fraction of energy states or levels occupied by electron among a group of energy states or levels at an energy E is given by the Fermi distribution function

$$f(E) = 1/[1 + \exp[(E - E_F)/kT]$$

where E_F is known as the Fermi energy and is the energy at which half of the energy states or energy levels among the group of energy states or energy levels at E_F are occupied by electrons. This assumes that there is a group of allowed energy states or levels at E_F. Even if there is not, E_F is still a useful reference energy to help determine what fraction of the energy levels located at some energies above E_F are occupied by electrons, or below E_F not occupied by electrons. Note each energy level can be occupied by two electrons, one spin up and one spin down, that is, each energy level has two electron states, a spin-up and a spin-down state.

(b) Incorrect. As described in section 231 on page 173, the Fermi-Dirac distribution is derived statistically from a model of many energy states or levels and many electrons. Thus, it is meaningless to apply it to a single state or single level. However, in an pictorial demonstration, only one dot (one electron) and one line (one energy level that can be occupied by two electrons with opposite spin ↓↑) are drawn but they represent a large number of replicas clustered at the same energy E_F in a ΔE range.

(c) Correct. As explained in Section 231 on page 173, f(E) also denotes the probability that an electron state is occupied by one electron or an energy level occupied by two electrons. Thus, the probability that the energy level at E_F is occupied by two electrons or the probability that the energy state at E_F is occupied by one electron is 50%.

P231.6 Obtain the correction terms to fourth order in energy (or E^4) in the Boltzmann approximation by expanding the Fermi function in the Taylor series. $\exp(Z) = Z^0/0! + Z/1! + Z^2/2! + Z^3/3! + Z^4/4! + ... + Z^n/n! + ...$ where n = integer or 0. How do the absolute and percentage errors vary with energy?

Let us define $Z = (E-E_F)/k_BT$ then,

$$f(Z) = 1/[1 + \exp(Z)]$$
$$= \exp(-Z)/[1 + \exp(-Z)]$$
$$= \exp(-Z)/[1 + 1 - Z + Z^2/2 - Z^3/6 + Z^4/24 + ...]$$
$$\Delta f(Z) = f(Z) - \exp(-Z) = - f(Z)\exp(-Z)$$
$$\rightarrow - \exp(-2Z) \text{ for } Z >> 1$$

The fractional error is then

$$\Delta f(Z)/f(Z) = - \exp(-Z) \simeq - [1 - Z + Z^2/2 - Z^3/6 + Z^4/24]$$

P231.9 At what energy is the Fermi function equal to 1/2? Calculate and tabulate the occupation probability of the energy levels at $(E-E_F)/kT$ = −5, −4. −3, −2, −1, 0, +1, +2, +3, +4, +5 **(a)** by electrons, f(E), and **(b)** by holes, 1 − f(E), **(c)** tabulate also the probability, and **(d)** the percentage error of both electron and hole occupation factors if the Boltzmann approximation is made. Does the solution obtained in P231.5 agree with your table?

The Fermi function is

$f(E) = 1/\{1 + \exp[(E-E_F)/kT]\} = 1/2$ at the energy $E = E_F$.

Note, the fractional error is given by

$$[\exp(-Z)-f(Z)]/f(Z) = \exp(-Z) = \text{exact}.$$

$(E-E_F)/kT$	f(E)	$\exp[-(E-E_F)/kT]$	%error
+5	6.693E-3	6.738E–3	0.674
+4	1.799E-2	1.832E-2	1.831
+3	4.742E-2	4.978E–2	4.979
+2	0.119203	0.134335	13.534
+1	0.268941	0.367879	36.788
0	0.500000	1.000000	100.000
–1	0.731959	2.718281	271.828
–2	0.880797	7.38906	738.906
–3	0.952574	20.08553	2008
–4	0.982914	54.59815	5460
–5	0.993307	148.4131	14841

P233.8 For a crystal of finite size, it is found that there are 100 allowed electron energy levels at the energy E_1 in the conduction band and that the Fermi energy lies at E_1, i.e. $E_F=E_1$. How many electrons are there whose energy is E_F?

There are 100 allowed electron energy levels at the energy E_1 and hence twice as many electron states to account for the spin up and spin down states. When $E_F=E_1$, half of the states or energy levels are occupied by electrons, thus, there are 100 electrons occupying half of the 200 electron states at energy E_1.

P233.9 For a crystal of finite size, it is found that there are 100 allowed electron energy levels at energy E_F. Which of the following statements are correct and wrong, and why? **(a)** Each of the 100 levels is occupied by one electron. **(b)** 50 levels are empty and 50 levels are each occupied by two electrons. **(c)** The first 50 levels are empty and the last 50 levels are each occupied by two electrons. **(d)** The probability that each level is occupied by one electron is 0.5. **(e)** 100 states are occupied by one electron each and 100 states are empty.

There are 100 allowed electron energy levels at an energy E_F. Because each energy level can be occupied by two electrons with two opposite spin, there are 200 allowed electron energy states at the energy E_F. The true-false answer to the questions of this problem must be a probability one since statistical distribution is a probability prediction rather than certainty so are the answers. The reason given is the key.

(a) Each of the 100 levels is occupied by one electron is correct but not a likely situation. Some levels can be occupied by two electrons and others none, but with most levels occupied by one electron, therefore statement is most likely incorrect because of its deterministicity.

(b) Not likely as indicated in (a).

(c) Not likely as indicated in (a)

(d) Incorrect. It should be the probability that each level is occupied by two electrons is 0.5. The probability that each level is occupied by one electron is 1.0 again is incorrect since probability of 1.0 means deterministic or certainty and hence no longer probabilistic.

(e) Again not precisely correct unless the word probability is added in front as follows. The probability that 100 states are occupied ('by one electron each' is implied and hence redundant but explicit for clarity for beginners) and 100 states are empty.

These examples show that easy common conversation language and simple 'one-glance' graphical illustrations imply an understood deep physics basis of Pauli's exclusion principle of electrons and many-electron statistics, i.e. quantum statistical mechanics.

P233.11 The electron concentration, $N=N_C\exp[-(E_C-E_F)/kT]$, and hole concentration, $P=N_V\exp[-(E_F-E_C)/kT]$, were derived based on four assumptions. What are these assumptions? Why are they made? And how are they justifiable? Partial answers: large crystal, spherical energy band, large band width, Boltzmann.

(1) Large crystal

The large crystal assumption was made so that the semiconductor crystal contains many closely spaced energy levels and the summation of the electrons occupying the electron energy states or levels in the conduction band may be replaced by an integral. This is justifiable since in 1 cm^3 of silicon, there are $N=(6.02\times10^{23}$ atoms/mole)(2.33 g/cm^3)/(28.0855 g/mole) = 4.99×10^{22} atoms and there are $2N=1\times10^{23}$ energy levels or $4N=2\times10^{23}$ energy states in the conduction band.

(2) Spherical energy band

As explained on page 176 in Section 233, the spherical energy band assumption, i.e. constant effective mass, is made to obtain a simple analytical expression for the density of states and N_C. This is an approximation of the ellipsoidal shaped constant energy surface of electrons in silicon which is a good approximation for electrons with kinetic energies near the band edge, E_C. A more complex expression for N_C is obtained if the ellipsoidal and multivalley three-dimensional (in the energy versus k-vector space) constant energy surfaces are taken into account but they will only change the effective mass in N_C to a more complex expression while the simple Boltzmann or exponential energy dependence of the electron and hole concentrations remains.

(3) Large conduction band width (>> kT)

The large band width approximation is made to simplify the evaluation of the integral by extending the upper limit of integration over energy to infinity, i.e., from ($E=E_C$ to E_C') to ($E-E_C=0$ to ∞), so that the resulting definite integral (with integration limits 0 and ∞) has an analytical solution in terms of known or tabulated functions. This is justifiable since the product of the occupancy factor and the density of states, f(E)D(E), peaks at an energy close to kT, and drops to negligibly small values at high energies (i.e. very few electrons with KE much higher than kT) due to the vanishingly small f(E) as $E \rightarrow \infty$ or $E-E_C$ becomes >> kT. So, the condition is $(E_{C'}-E_C)/kT >> 1$ or the conduction bandwidth is large compared with kT. A 4kT(≃0.1V at 300K) or larger bandwidth is sufficient. However, there are many important electrical conduction phenomenon due to the higher energy electrons with kinetic energies of several electron-volts (or equivalent temperatures of 40 times the room temperature, i.e. 12,000K therefore known as hot electrons, as estimated using 1eV ~ 40kT at 300K or kT≃25meV) which controls the specific performance of a p/n junction diode or a transistor or their reliability as measured by the operation time to failure (TTF_{op}). Therefore, the higher energy electrons (and holes) and the details of the energy band at high kinetic energies cannot be neglected except at or near thermal equilibrium.

(4) Boltzmann distribution of electron number versus electron kinetic energy

The use of the Boltzmann approximation for the Fermi-Dirac distribution again simplifies the evaluation of the integral used to calculate the electron and hole concentrations. This is justifiable at low electron concentrations (known as nondegenerate electron or hole density range) when the electron concentration, N, is small compared with the effective density of states of the conduction band, N_C, which is about $10^{18}cm^{-3}$. Then, the Fermi distribution function can be accurately approximated by the Boltzmann distribution function.

P233.12 Using the power series expansion for the Fermi function $f(X) = 1/(1+X) = 1 - X + X^2 - X^3 + X^4$ where $X = \exp[(E-E_F)/k_BT] = \exp[(E-E_C+E_C-E_F)/k_BT]$, obtain the correction terms to fourth order in Fermi energy in the bracket $\{1 + ...\}$ of the electron concentration expression $N = N_C\exp[-(E_C-E_F)/k_BT]\bullet\{1 + F_1Z + F_2Z^2 + F_3Z^3 + F_4Z^4\}$ where $Z=\exp[-(E_C-E_F)/k_BT] < 1$. Compute the Fermi energy position above which the electron concentration from the Boltzmann approximation is in error by 0.01%, 0.1%, 1%, and 10%.

This is an intermediate level problem illustrating the procedure used to develop accurate analytical approximation formulae to take into account high electron concentrations which are increasingly encountered in transistor design applications as the transistor dimension decreases and the impurity and carrier concentrations must be increased to maintain low statistical fluctuation in order to assure high manufacturing yield. The problem statement in FSSE has obvious typographic errors, which are corrected above, although the correct solution for the coefficients were obtained when the problem was initially designed and tested for construction of the curves in Fig.252.1 and 252.2.

The coefficients can be obtained from using the power series expansion given in the problem for the Fermi distribution function in the electron concentration integral given by (233.7) and integrate each term of the definite integral

$$\int\sqrt{x}\bullet\Sigma(n=1,5)(-1)^{n-1}\exp[-n(x_F+x)]dx$$
$$= \Sigma(n=1,5)(-1)^{n-1}(1/2n)(\pi/n)^{1/2}\exp(-nx_F)$$
$$= \Sigma)n=1,5)(-1)^{n-1}(1/2n)(\pi/n)^{1/2}Z^n$$

where we have used the definite integral with limits 0 and ∞ given by

$$\int\sqrt{x}\bullet\exp(-nx)dx = (1/2n)(\pi/n)^{1/2} = \Gamma(3/2)/n^{3/2},$$

where $\Gamma(m)$ is the Gamma function, the definition

$$Z \equiv \exp(-x_F) \equiv \exp[-(E_C-E_F)/k_BT],$$

and the Fermi energy which is measured relative to the conduction band edge and normalized to the thermal energy k_BT

$$x_F=(E_C-E_F)/k_BT.$$

The first term gives the usual N_C. Thus, the coefficients F_n in the correction factor

$$F_{correction} = (F_0 + F_1Z + F_2Z^2 + F_3Z^3 + F_4Z^3)$$

are given below where m=1 to 5

$F_{m-1} = (1/m\sqrt{m})$
$F_0 = +1/1\sqrt{1} = +1.0000000$
$F_1 = -1/2\sqrt{2} = -0.3535533$
$F_2 = +1/3\sqrt{3} = +0.1924500$
$F_3 = -1/4\sqrt{4} = -0.1250000$
$F_4 = +1/5\sqrt{5} = +0.0894427$

The electron concentration with the corrections terms is then

$N = N_C \cdot Z \cdot (F_0 + F_1Z + F_2Z^2 + F_3Z^3 + F_4Z^4).$

The second part of the problem on computing the Fermi level relative to the conduction band edge for 0.01% to 10% error in N can be obtained from solving the fourth order algebra equation given by the formula below and the numbers are for the students to work out. N/N_C = percentage error = $100Z(F_0 + F_1Z + F_2Z^2 + F_3Z^3 + F_4Z^4)$

P241.1 The experimental formulae for the intrinsic carrier concentration in silicon was obtained by Morin and Maita at the Bell Telephone Laboratories in 1954 from conductivity and Hall voltage measurements over a wide range of temperatures in many n-type and p-type impurity-doped silicon single crystals. Their results were fitted to give $n_i^2 = 1.5\times10^{33}T^3\exp(-1.21/k_BT)$ cm^{-6} where k_B is the Boltzmann constant given by (231.3) (k=8.616×10^{-5} eV/K, verify k in this unit) and T is the absolute temperature in Kelvin. Calculate the value of n_i at T=300C and verify your result with that given by Fig.241.1(a).

At T = 300K = 300K - 273.15 = 26.85C

$n_i^2 = 1.5\times10^{33}(300)^3\exp(-1.21/8.616\times10^{-5}\times300) = 1.893233\times10^{20}\text{cm}^{-6}$
$n_i = 1.376\times10^{10}\text{cm}^{-3}$. Fig.241.1(a) gives ~ $1.4\times10^{10}\text{cm}^{-3}$

At T = 300C = 300 +273.15 = 573.15K

$n_i^2 = 1.5\times10^{33}(573.15)^3\exp(-1.21/8.616\times10^{-5}\times573.15) = 6.46\times10^{30}\text{cm}^{-6}$
$n_i = 2.54\times10^{15}\text{cm}^{-3}$. Fig.241.1(a) gives ~ $2.4\times10^{15}\text{cm}^{-3}$

P242.5 What is the equilibrium concentration of electrons and holes in an n-type silicon at T=21.2°C (n_i=1.0x10^{10} cm^{-3}) in which N_{DD}=4x10^{10} cm^{-3} and all ionized, and N_{AA}=0? What is the error if the intrinsic electrons are neglected?

Use the (1) mass action law and (2) charge neutrality.
(1) $NP = n_i^2$ (2) $\rho = q(P - N + N_{DD} - N_{AA}) = 0$. Thus, from (1), we get $P = n_i^2/N$. Substituting into (2), we get $N^2 - N_{DD}N - n_i^2 = 0$. Then, solve for N noting that negative particle density is meaningless.

$$N = (N_{DD} + \surd(N_{DD}^2 + 4n_i^2))/2 = 4.236\times10^{10}\text{cm}^{-3}$$
$$P = n_i^2/N = 2.361\times10^9\text{cm}^{-3}.$$

If n_i is neglected, we would get a smaller electron concentration when we approximate it by N_{DD}, or $N' = 4\times10^{10}\text{cm}^{-3}$. This introduces a % error of $100\times(N' - N)/N = -5.6\%$. Therefore, n_i cannot be neglected in this case.

However, one cannot just add n_i to N'. This would give

$$N_{DD}+n_i = 4.0\times10^{10} + 1.0\times10^{10} = 5\times10^{10}\text{cm}^{-3}$$

which is larger than the correct value of $4.236\times10^{10}\text{cm}^{-3}$. The difference,

$$N_{DD}+n_i-N = (4+1-4.236)\times10^{10} = 0.764\times10^{10}\text{cm}^{-3}$$

is larger than the exact hole concentration, $0.236\times10^{10}\text{cm}^{-3}$. The fundamental cause is the nonlinear (or quadratic) electron-hole reaction kinetics, $N + P \leftrightarrow n_i^2$, or the nonlinear mass-action law $NP = K_i = n_i^2$, which makes the linear approximation, tacitly assumed in the addition exercise just made, invalid and in serious error.

P243.1 A homogeneously impurity doped silicon has $N_{DD}=1.01\times10^{16}$ phosphorus/cm^3 and $N_{AA}=1.00\times10^{14}$ boron/cm^3, all of which are ionized. What are the electron and hole concentrations at 21.2°C and 362°C? Use Fig. 241.1(a) or the experimental formula given in problem 241.1 to obtain the accurate value of n_i at these two temperatures.

(A) $T = 21.2°C = 21.2 + 273.15 = 294.35K$

$$n_i^2 = 1.5\times10^{33}T^3\exp(-E_G/kT)$$
$$= 1.5\times10^{33}(294.35)^3\exp[-1.21eV/(8.616\times10^{-5}eV/K)(294.35K)]$$
$$= (8.533\times10^9\text{cm}^{-3})^2.$$
$$n_i = 8.533\times10^9\text{cm}^{-3}.$$

From (1) mass action law and (2) charge neutrality: (1) $NP = n_i^2$

(2) $\rho = q(P - N + N_{DD} - N_{AA}) = 0$, we get

$$N = \{(N_{DD}-N_{AA}) + \surd[(N_{DD}-N_{AA})^2+4n_i^2]\}/2 = 1.000\times10^{16}\text{cm}^{-3}.$$
$$P = n_i^2/N = 7.281\times10^3\text{cm}^{-3}.$$

Note that $n_i \ll N_{DD} - N_{AA}$ so $N \simeq N_{DD} - N_{AA}$ is a very accurate approximation.

(B) $T = 362°C = 362 + 273.15 = 635.15K$

$n_i^2 = 1.5\times10^{33}T^3\exp(-E_G/kT)$

$= 1.5\times10^{33}(635.15)^3\exp\{-1.21eV/[(8.616\times10^{-5}eV/K)(635.15)]\}$

$= (9.797\times10^{15}cm^{-3})^2.$

$n_i = 9.797\times10^{15}cm^{-3}$. Use formulas in (A) above,

$N = \{(N_{DD}-N_{AA}) + \sqrt{[(N_{DD}-N_{AA})^2+4n_i^2]}\}/2 = 1.600\times10^{16}cm^{-3}.$

$P = n_i^2/N = 5.998\times10^{15}cm^{-3}.$

P244.1 What is the intrinsic temperature of an n-type Si with $N_{DD}=10^{18}cm^{-3}$.

Assume that all the donors are ionized. Then, from the definition of intrinsic temperature, $N=n_i(T)$, we have

$N_{DD} = 10^{18}cm^{-3} = N \equiv n_i = [1.5\times10^{33}T^3\exp(-1.21/kT)]^{1/2}\ cm^{-3}$

$\therefore (10^{18})^2/1.5\times10^{33} = 1000/1.5 = T^3\exp(-1.21/kT)$

$\therefore 1.21/kT = \log_e(T^3\cdot1.5/1000) = 3\cdot\log_eT - \log_e(1000/1.5)$

$\therefore T = (1.21/k)/[3\cdot\log_eT - \log_e(1000/1.5)]$ (Use $k=8.616\times10^{-5}eV/K$)

$= 1.404\times10^4/(3\times\log_eT - 6.50229)$

The above implicit equation can be solved iteratively to get T. Instead, we find the temperature to give $n_i=10^{18}cm^{-3}$ from Fig.241.1(a) which is T = 680C. Use this value, T=680+273.15=953.15K in the above equation, we get

$T = 1.404\times10^4/(3\times\log_e953.15 - 6.50229) = 997.37K = 724.2C$

The two values do not agree exactly but are close. The disagreement is that the formula is a least squares fit of experimental n_i vs T data below about 600K and it does not take into account the energy gap reduction with increasing temperature properly while the computed n_i-T curve in Fig.241.1(a) does, hence, the corrected show a lower temperature, 953.15K, to give $n_i=10^{18}cm^{-3}$ due to smaller energy gap, than the uncorrected formula, 997.37K, which neglected the energy gap drop at higher temperatures.

P252.1 A representative volume element dxdydz of a uniformly doped n-type silicon crystal contains 100 substitutional phosphorus donors distributed on the superlattice so there is no local fluctuation of their location. Which of the following statements are correct and incorrect? Give the fundamental reasons. **(a)** f_D is the fraction of the 100 donors each with one trapped electron. **(b)** f_D is the fraction of neutral donor. **(c)** $1-f_D$ is the fraction of the 100 donors unoccupied by electron or each without a bound electron. **(d)** $1-f_D$ is the fraction of donors each with one trapped hole. **(e)** $1-f_D$ is the ionized donor. **(f)** When $E_F=E_D$, half of the donors are neutral. **(g)** When $E_F = E_D - k_BT \cdot \log_e g_D$, half of the donors are ionized. **(h)** When $f_D = 1/2$, the first 50 donors are each occupied by an electron (or have each trapped or captured an electron), **(i)** When $f_D=1/2$, 50 of the 100 donors are each occupied by a spin-up electron. **(j)** When $f_D=1/2$, 50 of the 100 donors are each occupied by an electron, some have spin up and others have spin down. **(k)** When $f_D=1/2$, about 25 donors are each occupied by a spin-up electron and the other 25, a spin-down electron. **(l)** When $f_D=1/4$, 25 donors are each occupied by an electron and 75 donors are each occupied by a hole.

The key difference between this problem and P233.9 is on the occupation of the energy level of a bound-state localized at an impurity and a band-state spread out over the entire crystal. The bound-state energy level can only be occupied by one electron, but in one of two ways, spin-up or spin-down. Once the energy level is occupied by one electron, it no longer exists. The band-state energy level can be occupied by two electrons with opposite spin. We will assume that the probability and statistical thermal equilibrium concepts are understood and given by the fraction of donor states occupied by electrons, (252.3), $f_D = 1/\{1 + (1/g_D)\exp[(E_D-E_F)/k_BT]\}$ We shall focus on the ground state of the hydrogen model (1s bound state) for the phosphorus donor and consider only the spin degeneracy and not the configuration degeneracy due to the many energy minimum in the E-k diagram of Si (6 minimum along the six 100 directions at about $k_x a/2\pi = 0.8$ near the Brillouin zone boundary for Si conduction band shown in Fig.183.1).

For the 100 substitutional phosphorus donor atoms, there are 100 energy levels and 200 bound electron states with spin-up or spin-down, one level or two spin-states each localized at a phosphorus donor atom.

(a) Correct. **(b)** Correct. **(c)** Correct.

(d) Incorrect because there is no hole bound state so holes cannot be trapped at the phosphorus donor atom. **(e)** Correct.

(f) Incorrect because spin degeneracy or $g_D=2$ so $f_D=1/\{1+g_D^{-1}\}=2/3>1/2$

(g) Correct. This corrects the statement (f).

(h) Too deterministic from using the word 'first'. Ok without 'the first'.

(i) Not likely since this means a spin correlation of unity for the 50 trapped electrons which is most unlike due to statistical randomness. See (j).

(j) Correct but lacking quantitation from the word 'some'. See **(k)** Correct.

(l) Incorrect since there is no hole bound states at the phosphorus donor. The correct statement is '75 donors are not occupied'.

The perennial error of teachers and students alike has been the assumption of a hole bound state stated in two of the above questions, (d) and (l), and the confused algebra and erroneous formula obtained by using $1-f_D = 1/\{1 + g_D\exp[-(E_F-E_D)/k_BT]$ with energy sign changed to represent the occupation of the acceptor impurity energy level by hole. Instead, the powerful concept of symmetry of the quasi-particle electron and hole has not been exploited, which would have given directly (252.4) from (252.3) without any algebra. However, a further complication arises from the acceptor and donor terminology, from which the term accept seemed to have implied bound or capture into a quantum mechanical bound state. But in fact the term accept descript the accepting of an electron to fill up the empty valence bond from the lack of one (or more) valence electrons of the acceptor atom in comparison with the host atom. The less confusing descriptives are: electron trap and hole trap, which unambiguously describe the presence of a electron or a hole bound state. (See following solution for P252.3.)

P252.3 Give the fundamental reasons on the symmetry between the trapped electron and trapped hole distribution functions given by (252.3) and (252.4).

$$f_D = 1/\{1 + (1/g_D)\exp[(E_D-E_F)/k_BT]\} \qquad (252.3)$$

$$f_A = 1/\{1 + (1/g_A)\exp[(E_F-E_A)/k_BT]\} \qquad (252.3)$$

These two functions are identical when one recognizes that the electron kinetic and total energy is measured upwards and the hole kinetic and total energy is measured downwards. Thus, f_A would have had exactly the same form as f_D if expressed in terms of the hole kinetic and total energies. It is unwieldy to use two different sign conventions for energy, one for electrons and one for holes, in a problem where both electrons and holes, in the conduction and valence bands and bound to the electron and hole traps respectively are simultaneously present in the semiconductor. Thus, to follow the convention, that for electrons, the sign of the hole kinetic and total energies are changed to use the up direction for positive energy (for electrons). This sign change then interchanges the E_F-E_A terms to give that shown in f_A (252.4).

Chapter 3
DRIFT, DIFFUSION, GENERATION, RECOMBINATION, TRAPPING AND TUNNELING
(of Electrons and Holes)

OBJECTIVES

*** Understand basic concept: Two collision mechanisms.**

Scattering Collision

Trapping Collision

*** Know basics and usage of semiconductor parameters**

Drift Mobility and Velocity

Drift Velocity Saturation

Diffusivity (Diffusion Constant)

Thermal Generation Rate

Thermal Recombination Lifetime

Optical Generation Rate

Photon Absorption Coefficient

Interband Impact Generation Rate

Interband Auger Generation Rate

Interband Auger Recombination Rate

P310.1 Derive 3-d drift current density as a function of drift velocity.

Consider the 1-d drift current density in the x-, y-, and z-directions. The volume element used to calculate the drift current is shown on the right.

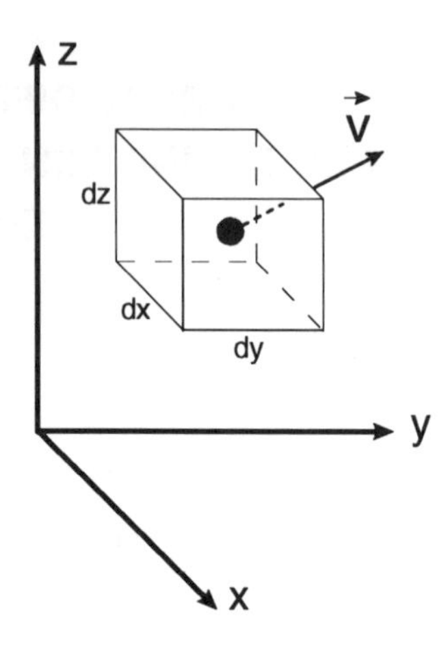

I_x = (electron charge)×(electron flux in #/cm²-sec)×(cross-sectional area) where (electron flux) = (electron volume density #/cm³) × (electron velocity cm/sec). Thus,

$$I_x = -q \times n \times (-dx/dt) \times (dydz)$$
$$= q \times n \times v_x \times dydz$$

where $v_x = -(dx/dt)$ is the component of the drift velocity in the x-direction. The negative sign shows that the electrons are drifting in the negative x-direction.

The drift current due to the drifting electrons is in the positive x-direction in the universal convention of denoting the direction of electrical current as the direction of motion of positively charged particles. Student should repeat this problem for holes to exercise your understanding of the principles underlying the two signs. Which direction is the hole drifting? This is very important because in transistors the current is composed of both the positively charged holes and negatively charged electrons which confuse the beginners when trying to relate the current direction with the direction of movement or drift of the electrons and the holes.

The electron current density (Ampere per unit area, A/cm^2 or A/m^2) is then

$$J_x = I_x/(dydz) = qnv_x$$

Similarly, in the y- and z-directions,

$$J_y = I_y/(dxdz) = qnv_y$$
$$J_z = I_z/(dxdy) = qnv_z$$

The 3-d drift current density is then given by the vector consisting of the x-, y-, and z-components derived above.

$$J = J_x i_x + J_y i_y + J_z i_z = qn(v_x i_x + v_y i_y + v_z i_z) = qn\mathbf{v}$$

P311.1 Verify that the rms noise in the drift velocity is $v_{dnoise} = (q/m)E_x\sqrt{[\langle\tau_f^2\rangle - \tau^2]}$. Show that because it is much smaller than the drift velocity, $v_d=(q/m)E_x\tau$, hence, it is even more smaller than the thermal velocity, $v_{th}=\sqrt{\langle v_o^2\rangle}=\sqrt{(3kT/m)}$. (As described on p.234-235 and (311.3) in FSSE, τ_f is used to denote the time of free flight of the electron, or hole, between the two successive collision events.)

Given: $v_d = (q/m)E_x\langle\tau_f\rangle = (q/m)E_x\tau$. Then, the rms noise is

$$
\begin{aligned}
v_d(\text{noise}) &= \sqrt{\langle[\;v_d - \langle v_d\rangle\;]^2\rangle} \\
&= \sqrt{\langle[(q/m)E_x\tau_f - (q/m)E_x\langle\tau_f\rangle]^2\rangle} \\
&= (q/m)E_x\sqrt{\langle[\tau_f - \tau\;]^2\rangle} \\
&= (q/m)E_x\sqrt{\langle[\tau_f^2 - 2\tau_f\tau + \tau^2]\;\rangle} \\
&= (q/m)E_x\sqrt{[\langle\tau_f^2\rangle - 2\langle\tau_f\rangle\tau + \tau^2]} \\
&= (q/m)E_x\sqrt{[\langle\tau_f^2\rangle - 2\tau^2 + \tau^2]} \\
&= (q/m)E_x\sqrt{[\langle\tau_f^2\rangle - \tau^2\;]} \\
&\equiv (q/m)E_x\sqrt{[\langle\tau_f^2\rangle - \langle\tau_f\rangle^2\;]}
\end{aligned}
$$

Therefore,

$$
\begin{aligned}
v_d(\text{noise}) &= (q/m)E_x\sqrt{[\langle\tau_f^2\rangle - \tau^2]} \\
&\ll v_d = (q/m)E_x\tau \\
&\ll v_{th} = \sqrt{(3kT/m)}
\end{aligned}
$$

because $\tau \simeq 10^{-13}$ sec. Thus,

$$v_d(\text{noise}) \ll v_d \sim 10^3\text{cm/s} \ll v_{th} \sim 10^7\text{cm/s}.\quad \text{Q.E.D.}$$

P312.1 Describe the conditions at which the conductivity is an equilibrium parameter and hence a fundamental property of a solid.

The conductivity is equal to the product of the electronic charge, q, electron or hole mobility, and electron or hole concentration, $\sigma_n = q\mu_n n$ and $\sigma_p = q\mu_p p$. Therefore, if the carrier mobility and concentration are measured at thermal or thermodynamic equilibrium, then the conductivity is an equilibrium characteristic or parameter and hence a fundamental property of the solid measured which is independent of the measurement condition or the measuring force applied. To make the conductivity measured independent of the measuring or applied force requires that the applied force (such as a d.c. voltage) be as small as possible while the response (such as the d.c. current) to the applied force can still be measured above the background noise.

The following is a reiteration of what were described in section 202 on the conditions of equilibrium. The carrier concentration and mobility have their equilibrium or fundamental values when there is no locally or spatially inhomogeneous applied d.c. or a.c. excitations or forces (i.e. no localized mechanical force, and no localized electromagnetic field and no localized or focused light or photons), i.e., it is at thermodynamic equilibrium with its ambient (or heat sink) by means of energy exchange via collisions of ambient molecules with the solid's surface atoms and the inter-atom collisions within the solid to make the average kinetic energy of each particle species equal (or simply but abstractly stated, space-time constant temperature).

P312.2 Is the definition of conductivity limited to thermal equilibrium?

No. The conductivity may also be measured under nonequilibrium conditions, but then it is a function of the applied excitation (such as the amplitude and frequency or waveform of the applied excitation). Thus, the conductivity measured at nonequilibrium is NOT a fundamental property of the solid because it depends on some properties of the applied excitation. This is a very important consideration for determining experimentally the fundamental properties of a semiconductor and basic characteristics of a transistor. In fact the first and simplest model of transistor to be learned in this course is a linear model derived from infinitesimal applied force, however, the requirement of an infinitesimal applied force has seldom been stated.

P312.3 Is the definition of conductivity limited to a homogeneous solid?

No. The definition of conductivity is easily extended to inhomogeneous solids with spatially varying electron or hole concentrations and mobilities. The conductivity of an inhomogeneous solid, measured between two leads in contact with the solid is then the integrated conductivity over the volume. It is just like a network connection of many resistors (each has two leads) of different values in parallel and in series. To compute the resistance between the two terminals or any two nodes of this resistance network, one combines the parallel and series resistance formulas. The difference is size: the resistors are each a large body whose resistivity or conductivity can be inhomogeneous but treated as a lump known as a lump element while the inhomogenous regions in a solid (known as grains or crystallites for the case of a polycrystalline solid) are small, macroscopic size known as a distributed element.

P312.4 Is the drift current formulae limited to a homogeneous solid?

The drift current formulae is not limited to a homogeneous solid and may be extended to a inhomogeneous solid with spatially varying conductivity. In an inhomogeneous solid, the drift current would be spatially varying due to the spatially varying conductivity. The spatially varying conductivity can come from the inhomogeneity of one or both of its two parts, the carrier concentration and mobility. If the carrier concentration varies spatially, there would be an additional current due to carrier diffusion in its concentration gradient, to be discussed in section 320.

P312.5 Is the drift current formulae an equilibrium or nonequilibrium expression?

The drift current formulae can be either an equilibrium or a nonequilibrium expression since the carrier concentration and electric field during measurement may be at either condition. [The remaining can be included after studying sections 32n.] At equilibrium, all particle currents are individually equal to zero, i.e. $\mathbf{J}_N=\mathbf{J}_P=0$. But the particle current is the sum of the drift and diffusion currents. Therefore, at thermal equilibrium, the drift current may not be zero, but it must be exactly canceled by the diffusion current. The diffusion current arises from a concentration gradient of the charge carriers (electrons or holes), for example, from a spatially varying impurity concentration. However, this is not strictly an equilibrium condition because impurity atoms will diffuse due to its concentration gradient but the impurity atom diffusion current is negligible at room temperatures. Thus, the solid is at a partial equilibrium state of electrical (electronic) equilibrium which was described in section 202 qualitatively and now demonstrated quantitatively with drift and diffusion currents.

P312.6 A current of 1 A/cm^2 is forced through an n-type Si bar with a resistivity of 1 ohm-cm at 300K. What is the drift velocity of the electrons? How long does it take for an electron to drift through a Si bar of 1mm long?

Start with Ohm's law, $V = IR = I\times(\rho L/A) = \rho L \times(I/A) = \rho LJ$ where ρ = resistivity = 1 ohm-cm, L = length = 1mm, and J= current density = 1 A/cm^2. Solve for the voltage drop across the n-type Si bar, which is

$$V = (1\ \text{ohm-cm})\times(0.1\ \text{cm})\times(1\ \text{A/cm}^2) = 0.1\ \text{V}.$$

The electric field is then given by

$$E_x = V/L = 0.1\ \text{V}/0.1\ \text{cm} = 1\ \text{V/cm} = 100\ \text{V/m}.$$<ps=12<

The drift velocity of the electron is calculated using (311.4), $v_d = (q/m)E_x\tau$ and the values given on p.236 of FSSE for the electronic charge, q, the electron mass, m, and the scattering mean free time, τ.

$$v_d = (1.60\times10^{-19}\text{C}/9.11\times10^{-31}\text{kg})\times(10^{-13}\text{s})\times(100\text{V/m}) = 1.75\text{m/s} = 175\text{cm/s}.$$

Integrating the velocity defined by, $v = dx/dt$, we get

$$x = x_0 + v\times(t - t_0).$$

Let $x_0 = 0$ and $t_0 = 0$. The time required for the electron to drift 1mm = 0.1cm is then given by, (ms=millisecond)

$$t = x/v_d = (0.1\text{cm})/(175\text{cm/s}) = 5.71\times10^{-4}\text{s} = 0.571\ \text{ms}.$$

P313.10 An Si crystal is uniformly doped by 1.000×10^{17} B/cm^3 and 1.001×10^{17} As/cm^3. Compute the conductivity using Fig.313.5 or the formulae in Table 313.1.

$$N_{DD} = 1.001\times10^{17}\ \text{As/cm}^3 \qquad N_{AA} = 1.000\times10^{17}\ \text{B/cm}^3$$

To calculate the majority carrier (electrons) concentration, use (242.7)

$$N = {}^1/_2\{N_{DD} - N_{AA} + \surd[(N_{DD} - N_{AA})^2 + 4n_i{}^2]\} \qquad (242.7)$$

Assume $n_i = 10^{10}\text{cm}^{-3}$ at T=296.5K which is the standard condition unless otherwise specified by the problem. Then use the given 'net' impurity concentration for this problem,

$$N_{DD} - N_{AA} = 1.001\times10^{17}\text{cm}^{-3} - 1.000\times10^{17}\text{cm}^{-3} = 1\times10^{14}\text{cm}^{-3}$$

the majority carrier or electron concentration is

$$N = (1/2)\{10^{14}\text{cm}^{-3} + \surd[(10^{14}\text{cm}^{-3})^2 + 4(10^{10}\text{cm}^{-3})^2]\} = 1\times10^{14}\text{cm}^{-3}$$

which is equal to the net impurity concentration, $N_{DD} - N_{AA}$, within truncation error. From the mass action law, the minority carrier (holes) concentration is

$$P = n_i{}^2/N = (10^{10}\text{cm}^{-3})^2/10^{14}\text{cm}^{-3} = 1\times10^{6}\text{cm}^{-3}$$

which is much greater than the majority carrier (electron) concentration.

The extrinsic conductivity is given by (315.2),

$$\sigma = \sigma_n + \sigma_p = q\mu_n n + q\mu_p p \qquad (315.2)$$

Since N >> P, the extrinsic conductivity is due almost entirely due to the majority carriers (electrons),

$$\sigma = \sigma_n + \sigma_p \simeq \sigma_n = q\mu_n n.$$

The majority carrier mobility for electrons and holes are plotted in Fig.313.5 versus the concentration of ionized scatterers, N_I (cm^{-3}). The concentration of ionized scatterers is equal to the **sum** of the acceptor and donor impurity concentrations.

$$N_I = N_{AA} + N_{DD} = 1.000\times10^{17}\text{cm}^{-3} + 1.001\times10^{17}\text{cm}^{-3} = 2.001\times10^{17}\text{cm}^{-3}.$$

At $N_I = 2.001\times10^{17}\text{cm}^{-3}$, we find from Fig.313.5 approximately $\mu_n \simeq 500\text{cm}^2/\text{V-s}$. Therefore, the conductivity is given by

$$\sigma \simeq \sigma_n = q\mu_n n = (1.602\times10^{-19}\text{C})(500\text{cm}^2/\text{V-s})(10^{14}\text{cm}^{-3}) = 8.01\times10^{-3}\text{S/cm}.$$

Its reciprocal is the resistivity which is $\rho=1/\sigma=1/8.01\times10^{-3}\text{cm/S} = 125\text{ohm-cm}$.

The electron mobility may also be calculated using the Matthiessen rule and the equations for the lattice scattering limited and ionized impurity scattering limited mobilities tabulated in Table 313.1. This would give more significant figures than reading off Fig.313.5 but the accuracy is not necessarily better because of the experimental errors of the data used to obtain the mobility equations.

P331.1 A current of 1 A is forced through a homogeneous n-type 1 ohm-cm Si bar of cross-sectional area of 1 mm^2. What is the quasi-Fermi electric field of electrons in the Si bar? What is the quasi-Fermi electric field of holes in the Si bar? What are the electron and hole quasi-Fermi potential differences between the two ends of the bar of 1 cm? What is the applied voltage across the 1 cm bar? Are your answers consistent, why?

The solutions of this problem can be written down in a few lines of equations given below. However, the discussions given below are necessary to provide the physics underlying the assumptions which are used to derive the equations.

The total conduction current (diffusion and drift) is given by $J = J_N + J_P$. The electron current, J_N, and the hole current, J_P, are given in terms of the quasi-Fermi potential for electrons and quasi-Fermi potential for holes by (331.1A) and (331.3A). Let us orient the Si bar so that its length L of 1cm is along the x-axis and let us also assume that the 1A current enters the Si bar at x=0 and exits at x=L. (The magnitude of the numerical solution to be given is independent of orientation and which end the current enters the Si bar or current direction since the Si bar is homogeneous.) When not specified by the question, one dimensional analysis is assumed valid which is a good approximation if the silicon piece is large and the metal (or low resistance) contact covers each end completely and makes areally independent uniform contact to the Si so two-dimensional and three-dimensional geometrical distortions of the current flow or carrier (electrons and holes) particle flux are unimportant. Then the one dimensional current equation in terms of the quasi-Fermi potential is

$$J = q\mu_n N(-dV_N/dx) + q\mu_p P(-dV_P/dx) \equiv \sigma_n(-dV_N/dx) + \sigma_p(-dV_P/dx).$$

The Si bar is n-type, therefore the conductivity is nearly entirely due to electrons because the electron (or majority carrier) concentration is much larger than the hole (or minority carrier) concentration, $\sigma_n >> \sigma_p$ so $\sigma \simeq \sigma_n$ and

$$J = J_N + J_P \simeq J_N = q\mu_n N(-dV_N/dx).$$

Rearranging, the quasi-Fermi electric field for electrons is given by:

$$(dV_N/dx) = -\rho \times J = -\rho \times (I/A) = -(1\text{ohm-cm}) \times (1A)/(1\times 10^{-2}\text{cm}^2) = -100V/\text{cm}.$$

The negative sign comes from the choice we made on which end of the bar the (positive) current enters. (We assumed the x=0 end.)

Because the silicon bar is homogeneous (which means the impurity concentration is spatially constant) and there is no localized light or other localized excitation to generate excess electrons and holes in some parts of the silicon bar which is the standard condition if it is not explicitly stated otherwise in a problem (which means that carrier or electron and hole concentrations are also spatially constant), the silicon bar is near thermal equilibrium. (There is no localized heating either by the 1A current which would have caused a spatially varying temperature distribution that could cause a thermoelectric current.) But it is not at electrical equilibrium because of the 1A current. Because of thermal equilibrium and homogeneity, the quasi-Fermi energy level or quasi-Fermi potential for the electrons and holes are the same or coincide along the entire length of the silicon bar.

The quasi-Fermi potential difference between the two ends of the 1cm Si bar is found by integrating the quasi-Fermi electric field through the 1cm length. The model we took at the beginning of the answers assumed that the x-axis is alone the axial direction of the Si bar and the 1A (positive) current enters the Si bar at x=0, thus,

$$\Delta V_N = V_N(x=L) - V_N(x=0) = \int (dV_N/dx)dx = (dV_N/dx)\times L$$
$$= (-100V/\text{cm})\times(1\text{cm}) = -100V$$
$$\Delta V_P = \int (dV_P/dx)dx = (-100V/\text{cm})\times(1\text{cm}) = -100V$$

The voltage drop through the length of the bar is exactly the quasi-Fermi potential difference just computed. This can also be obtained alternatively by noting that this is a piece of homogeneous resistor with only drift current and no diffusion current (because there is no carrier concentration gradient) so that the electrostatic potential drop through the length of the Si bar is the same as the voltage drop and voltage drop is also the strictly correct quasi-Fermi potential drop and electrochemical potential drop. So, to recheck the consistency between electrostatic potential drop and quasi-Fermi potential drop, we use Ohm's law, and the voltage drop through (old-timers use 'across' instead of 'through' which is a misnomer that confuses beginners because the bar's across or lateral dimension is irrelevant or very large in this one-dimensional problem) the bar is given by $V = IR = I\times(\rho L/A) = \rho L\times(I/A)$ where ρ = resistivity = 1Ω-cm, L = length = 1cm, I = 1A, and A = area = $1mm^2 = 1\times10^{-2}cm^2$. Then, the potential drop is V = $(1\Omega\text{-cm})\times(1\text{cm})\times(1\text{A})/(1\times10^{-2}\text{cm}^2) = 100\text{V}$. By the traditional or IEEE sign convention, the voltage drop is the voltage at the (positive-charged-particle) current exit point minus the the voltage at the current entrant point, or $V(x=L) - V(x=0) = -100V$.

P381.1 An n-type Si crystal (10^{17} donor/cm^3) contains 10^{15} Au/cm^3. What are the low level and high level lifetimes of electrons and holes in this sample?

Gold has two energy levels (or electronic or electron and hole bound states) in the Si energy gap. Electron-hole recombination can occur at these two Au levels via the four band-to-trap and trap-to-band electron and hole transitions shown in Fig.3621.1 in which energy conservation is effected by lattice vibration to dissipate the recombination energy and to supply the generation energy. This is the thermal recombination and generation mechanism via an intermediate trap or bound state and known as the Shockley-Read-Hall process after its originators who first formulated the simple and concise model depicted in Fig.3621.1 The underlying physics and the trapping kinetics of recombination and generation of electrons and holes are reviewed in sections 362 and 372 respectively.

The solution of this problem requires two steps: (i) find a way to compute the values of the electron and hole capture rate coefficients at the gold energy levels in Si from related experimental data (like the thermal emission rates which are measured and shown in Fig.381.2) or get the values from experimental data (not given in the book and hard to find in the research journals), and (ii) compute the lifetimes using the appropriate formulas derived in the textbook.

Since the problem did not specify, we will assume the traditional and simplest situation, that of thermal equilibrium or near thermal equilibrium, so that the fundamental transition rate of electron capture by and electron emission from the gold trap, and similarly those of holes, are at their respective equilibrium values. This near-equilibrium condition is denoted by the subscript e (for equilibrium) in the thermal capture rate c^t_{ne} and thermal emission rate e^t_{ne}. They follow a simple equilibrium relation, the mass action law, $e^t_{ne}=c^t_{ne}n_1$ given by (381.4). Thus, the capture rate can be calculated from the experimental emission rate given in Fig.381.2 in order to compute the lifetime defined in (372.3), $\tau_n(SRH)=1/(c^t_{ne}N_{TT})$ where N_{TT} is the total concentration of the electron trap and c^t_{ne} is the equilibrium electron capture rate coefficient. Using $n_1=n_i\exp[(E_I-E_T)/kT]$ then

$$c^t_{ne} = e^t_{ne}/n_1 = e^t_{ne}/\{n_i\exp[(E_I-E_T)/kT]\}$$

Use the gold energy level near the midgap, $E_T-E_V = 590\text{meV}$ and $E_C-E_T = 547\text{meV}$, and the Si energy gap value used in the experimental data given in Fig.381.2 which is 590 + 547 = 1.14 eV. Then, the intrinsic Fermi level, E_I, is

$$E_I = (1/2)[(E_C + E_V) + kT\times\log_e(N_V/N_C)] \quad (241.2)$$

$$E_I - E_V = (1/2)[(E_C - E_V) + kT\times\log_e(N_V/N_C)]$$

where $E_C - E_V$ = energy gap = E_G. $N_V=1.30\times10^{19}\text{cm}^{-3}$, $N_C=2.75\times10^{19}\text{cm}^{-3}$ from p.179, and $kT=0.02585$ eV at T=300K. Then,

$$E_I - E_V = (1/2)[(1.14\text{ eV}) + (0.02585\text{eV})\log_e(1.30\times10^{19}/2.75\times10^{19})$$

$$= 0.5603\text{ eV} = 560\text{ meV}.$$

From E_T - E_V given above and E_I - E_V, then,

$$E_I - E_T = (E_I-E_V) - (E_T-E_V) = 560\text{meV} - 590\text{meV} = -30\text{meV} = -0.030\text{eV}.$$

Now, calculate $n_1=n_i\exp[(E_I-E_T)/kT]$ at 300K, using $n_i=10^{10}\text{cm}^{-3}$ and $kT=0.02585\text{eV} = 25.85\text{meV}$) we get

$$n_1 = (10^{10}\text{cm}^{-3})\exp(-30\text{meV}/25.85\text{meV}) = 0.3133\times10^{10}\text{cm}^{-3}.$$

From Fig.381.2, $1/e^t_{ne}\simeq10^{-3}$ second at T=300K so $e^t_{ne}\simeq1000\text{s}^{-1}$. Thus,

$$c^t_{ne} = e^t_{ne}/n_1 = (1000\text{s}^{-1})/(0.3133\times10^{10}\text{cm}^{-3}) = 3.2\times10^{-7}\text{cm}^3/\text{s}.$$

Similarly, for holes, $e^t_{pe}=c^t_{pe}p_1$ where

$$p_1 = n_i\times\exp[(E_T-E_I)/kT]$$

$$p_1 = (10^{10}\text{cm}^{-3})\times[\exp(30\text{meV}/25.85\text{meV})] = 3.1917\times10^{10}\text{cm}^{-3}.$$

From Figure 381.2, $1/e_p{}^t\cong10^{-2}$ second at T=300K so $e^t_{ne}\simeq100\text{s}^{-1}$. Thus,

$$c^t_{pe} = e^t_{pe}/p_1 = (100\text{s}^{-1})/(3.1917\times10^{10}\text{cm}^{-3}) = 3.133\times10^{-9}\text{ cm}_3/\text{s}.$$

With the equilibrium capture rate coefficients numerically computed as above, we can now compute the steady-state low injection level and high injection level Shockley-Read-Hall (SRH) lifetimes.

Low level lifetime:

$$\tau_n = 1/c^t_{ne}N_{TT} = 1/(3.2\times10^{-7})(10^{15}) = 3.1\times10^{-9}\text{s} = 3.1\text{ns} = \tau_{n0}.$$

$$\tau_p = 1/c^t_{pe}N_{TT} = 1/(3.1\times10^{-9})(10^{15}) = 3.1\times10^{-7}\text{s} = 310.\text{ ns} = \tau_{p0}.$$

High level lifetime:

$$\tau_n = \tau_p = \tau_{n0} + \tau_{p0} = 3.1\times10^{-9}\text{s} + 3.2\times10^{-7}\text{s} = 3.23\times10^{-7}\text{s} = 323\text{ns} \simeq \tau_{p0}.$$

Chapter 4
METAL-OXIDE-SEMICONDUCTOR CAPACITOR

OBJECTIVES

*** Fabrication Process Design**

- Oxidation
  - Temperature, Time, Heating-Cooling Rates$^+$
  - Ambient (O_2, O_2+H_2O+Ar+N_2O+HCl)
- Annealing
  - Temperature, Time, Heating-Cooling Rates
  - Ambient (O_2, H_2+N_2/forming gas)$^+$

  ($^+$ For interface and oxide trap control and reliability.)

*** Device Design**

- Oxide Thickness
- Gate Type
  - Metal (Al)
  - Polycrystalline Silicon (doped)
  - Refractory Metal Silicide (W, Ti, Ta, Mo, ...)
- Bulk Dopant Impurity Type
  - n-type (P,As), p-type (B)

*** Device Analysis**

- Charge Control
- Differential Equation
- Energy band Diagram Application
- Transients
- Small-Signal Admittance (CTSA)

*** Device Characterization**

- $C(V)$, $C_{HF}(V)$, $C_{LF}(V)$, $C_{Depletion}$, $C_{Qstatic}=I/(dV/dt)$, $I(V)_{Tunnelling}$
- $C(t)$, $I(t)$ (charge trapping transients)
- $Y(V,\omega) = G(V,\omega) + j\omega C(V,\omega)$

*** Material Characterization**

- Oxide Thickness
- Dopant Impurity Concentration Profiling
- Oxide and Interface Trap Density Profiling

*** Reliability**

- Oxide Integrity (pin holes and dielectric breakdown)
- Oxide and Interface Trap Generation and Annihilation

P400.1 Why is i=C(dV/dt) inconsistent with i=dQ/dt if Q=CV is used? Is C=dQ/dV consistent with i=C(dV/dt) and i=dQ/dt? Explain concisely, succinctly.

If Q=CV, then d=dQ/dt=d(CV)/dt = C(dV/dt) + V(dC/dt). This has the extra term V(dC/dt) which comes from a capacitance that can vary with applied voltage, V.

C=dQ/dV=(dQ/dt)/(dV/dt)=i/(dV/dt) is consistent with i=dQ/dt.

The problem encountered here and faced by many electrical engineers taught with linear circuit is that the capacitance is a nonlinear circuit element, such as the capacitance of a MOS capacitor. Its nonlinearity comes from the voltage dependence of the stored charge. Therefore, the capacitance defined by Q=CV or C=Q/V is the total static capacitance and it is not the same as the capacitance defined by C=dQ/dV which is the differential capacitance or the slope of the nonlinear Q versus V curve. Here, we have considered only the d.c. case, that is Q and V are d.c. quantities. This can be extended to high-frequencies which would introduce a third capacitance known as the small-signal capacitance which is a function of the applied d.c. voltage but also a function of the frequency of the small-signal sinusoidal voltage or current used to measure the capacitance, such as by an admittance or impedance bridge.

P401.1 The n-channel Si MOS transistor switch in the DRAM (dynamic random access memory) cell of the next generation 16Mbit chip (in 1991. Now 1996, 256Mbit. Note: 1 bit = 1 cell which contains 1 transistor and 1 capacitor.) requires an oxide of 100A or 10nm (1A = 10^{-8}cm = 0.1nm and 1nm = 10^{-9}m = 10^{-7}cm = 10A). The cell sizes or lithographic line width are about 0.7μm = 7000A = 700nm in order to pack 16 million cells on one 0.5cm×0.5cm chip. This small line width requires low temperature processing so that the high temperature oxidation and diffusion times are not so short to make the process not controllable. Suppose that the 100A thick oxide is to be grown at 900C in dry oxygen on a p-type Si.

(a) What is the oxidation time in seconds. (Answer: Near 300 seconds.)

Using the oxide thickness versus oxidation temperature curve labeled 760mm pressure of oxygen and no water and argon in Fig.401.2, at 900C, the oxidation rate is approximately $x_0/\sqrt{t_0}=0.035(\mu m/\sqrt{hour})$. Thus, for x_0 = 100A = $10^{-2}\mu m$ = 10^{-6}cm, the oxidation time is $t_0 = (x_0/0.035)^2 = (10^{-2}/0.035)^2 = 0.0816$ hour = 294 seconds.

(b) What is the capacitance of the oxide layer per unit area, C_o, in pF/cm^2?

$C_o = \varepsilon_o/x_o = [3.9\times8.854\times10^{-14}(F/cm)]/10^{-6}cm$
$= 3.45\times10^{-7}F/cm^2 = 3.45\times10^5 pF/cm^2$.

A more convenient unit is $3.45fF/\mu m^2$ where $1fF=10^{-15}F=10^3pF$ and $1cm=10^4\mu m$ since the area of the capacitor is measured in units of micrometers or μm's in the 1996 manufacturing technology.

(c) If this is used with the traditional 5V power supply, what must be the capacitor area in order that the stored electron number is not less than a quarter million to avoid error from random electrical noise, such as the thermal noise. What is the minimum total capacitance area on the 16MBit DRAM chip?

$Q = 0.25\times10^6 q = C_o A_{MOS} V$. Thus, the MOSC area is

$A_{MOSC} = 0.25\times10^6 q/C_oV = 0.25\times10^6\times1.602\times10^{-19}C/[3.45\times10^{-15}(F/\mu m^2)\times5V]$
$= 2.32\mu m^2$.

The minimum total capacitance area of the MOSC in the 16Mbit-DRAM is $2.32\times16\times10^6\mu m^2 = 0.37cm^2$. The actual chip area is roughly twice larger to include one on-off MOST switch per capacitor, the interconnection line area, and the refresh and multiplexor transistor circuits.

P402.1 The explanation given for the HFCV curve is that the minority carriers cannot be supplied to and extracted from the oxide/silicon interface rapidly enough by diffusion or generation-recombination to respond to the signal voltage variation at 1MHz. Why is recombination also necessary and not just generation?

Considering p-Si so electrons are the minority carrier. During the positive part of the small-signal sinusoidal voltage applied to the gate, electrons in the p-Si bulk ('bulk' was the buzz word for 'interior'.) region are attracted to the SiO_2/Si interface and they must be replenished by electron generation in the Si bulk. During the negative part of the small-signal gate voltage, electrons at the SiO_2/Si interface are pushed away from the interface into the silicon bulk and they must recombine with the holes in the p-Si bulk at recombination centers to maintain the original electron concentration at zero signal or the steady-state averaged electron concentration with the applied small-signal. Thus, to respond to the high signal frequency and to contribute to the measured capacitance as

$(dq_g/dt)/(dv_g/dt)$ or Q_g/V_g where Q_g and V_g are the sinusoidal small-signal rms value, the generation and recombination rates must be sufficiently high. Usually, the rates are small and the recombination-generation events cannot keep up with the sinusoidal voltage variation at 1 MHz. Therefore, the measured capacitance at this frequency does not contain the minority carrier contribution, only the majority carrier contribution.

P410.1 Verify the negative sign in $C_s = -dQ_S/dV_S$ and $C_{it} = -dQ_{IT}/dV_S$ via physical argument without algebra by noting that increasing V_S more positively would increase N and decrease P.

The answer must be independent of the substrate type, n-Si or p-Si, and independent of orientation or coordinate choice of the capacitor. It does depend on the location choice of the reference for the electric potential, such as $V(0) = V_S$ and $V(x=\infty)=0$, which indeed can serve as a check based on the current and electron-hole particle or charge carrier flow directions. Thus, as V_S increases or becomes more positive, the Si electron energy band at the SiO_2/Si interface bends downward more, and more electrons from the Si bulk (n-type or p-type does not matter) are attracted to the interface. This increases the number of electrons stored in the thin Si surface space-charge layer next to the SiO_2/Si interface which is represented by the differential charge-storage capacitance, C_s, thus a negative sign is needed to compensate for the negative sign of the electron charge in order to give the capacitance convention of an increase of positive charge by an increase of applied positive voltage to the capacitor. Similarly, the larger electron concentration at the interface means a larger number of electrons will be trapped by the interface traps or the interface trapped charge density, Q_{IT}, will be more negative. Thus, a negative sign is again needed to represent the differential increase of the number of trapped electrons from a differential increase of the applied positive voltage by a positive charge storage capacitance of the interface traps, C_{it}.

If we had used the x=0 interface plane for the electric potential reference $V(x=0)=0$, and let $V(x=\infty)=-V_S$, then the negative sign would not have appeared. This illustrates the origin of the negative signs. This alternative reference is inconvenient, even confusing, because of its finite potential at infinity and its direction just opposite to the IEEE convention of left-to-right for input-to-output for transistor circuits.

P410.3 The integral in Problem P410.2 has been used by some researchers to compute the interface trap density-of-states in order to monitor an oxidation or a whole production process. Discuss the limitations of applying the above results to experimental C_g-V_G to compute the density of the interface traps given by C_{it} defined by (410.13) such as the uncertainty of the energy level position of the interface traps.

There are two uncertainties in the integral, the C_o and the reference potential for V_S. The error in C_o would give a very large error in the computed C_{it} or D_{IT} because it is obtained by taking the derivative of the experimental C_g-V_G curve with respect to V_G. The inability to determine the reference potential for V_S prevents a determination of the computed D_{IT} as a function of interface trap energy in the silicon energy gap. Another experimental method must be used to give the reference potential.

P411.2 The thickness ratio of the inversion layer, x_i, to the strong-inversion space-charge layer, x_d, was quoted on p.328 as $x_i/d_d = 0.414/1.414 = [(\sqrt{2}-1)/\sqrt{2}]=0.414/1.414=1/3.415$ while discussing the space-charge distribution shown in Fig.410.1(b). Show how this ratio can be obtained from the depletion approximation. (The original problem used x_{th} instead of x_d which was a typo.)

There are two solutions. The inversion and strong inversion conditions are defined by the minority carrier concentration at the SiO_2/Si interface x=0: $N(x=0)=P(x=0)=n_i$ for inversion or intrinsic surface, and $N(x=0)=P(x=\infty)\simeq N_{AA}$ (for p-Si) for strong inversion or 'on-set' of very-strong inversion. The corresponding surface potentials or total energy band bendings in the semiconductor are: $V_I(0)=V_F$ for the intrinsic surface and and $V_I(0)=2V_F$ for the strongly inverted surface. Thus, using (411.3A), $x_d=(2\varepsilon_S V_S/qN_{AA})^{1/2}$, then $x_d(\text{strong-inversion}) \simeq x_d(V_S=2V_F)$ and $x_i=x_d(V_S=V_F)$ so x_i/x_d is $1/\sqrt{2} = 0.707$. This answer is for two different applied gate voltages, one at $V_G=V_{GI}$ and the second at $V_G=V_{GT}$ or $V_G>V_{GT}$.

The thickness ratio of the inversion and strong inversion layer at one applied voltage, sufficiently large to cause strong inversion is $x_i/x_d = [x_d(V_S=2V_F) - x_d(V=V_F)]/x_d(V_S=2V_F) = (\sqrt{2}-1)/\sqrt{2} = 0.414/1.414 = 1/3.415$ which is substantially smaller than that given by the two-applied-voltage comparison just made above. Here, x_i is the thickness of the inversion channel from x=0 to x_i with the potential reference $V(x_d)=0$ and $V(x_i)=V_F$ where $N(x_i)=P(x_i)=n_i$.

P411.3 The MOSC area in a DRAM cell is $A=10\mu m^2$ and it is fabricated on an n-type Si substrate with a donor dopant concentration of $N_{DD} = 1.0\times10^{17}cm^{-3}$. The manufacturing technology is so refined and under control that there are no oxide and interface traps and that the work function difference between the gate conductor metal and the Si is zero by properly doping the polycrystalline silicon gate conductor. The chip is at room temperature. Assume T=290K, kT/q = 0.025V, and $n_i=1.0\text{x}10^{10}cm^{-3}$.

(a) What is the oxide capacitance in pF and fF units? (ans. 50fF)
(1pF=10^{-12}Farad and 1fF=10^{-15}Farad). $\varepsilon_{ox}=3.9\times8.854\times10^{-14}$F/cm.

We let X_o=69.06A to make C_O=50fF given by problem.

$$C_0 = C_oA \equiv \varepsilon_0 A/X_o = 50\text{fF (given answer)} \qquad \text{(P411.3A)}$$
$$= (3.9\times8.854\times10^{-14}\text{F/cm})(10\mu m^2)(10^{-4}\text{cm}/\mu m)^2/(69.06\times10^{-8}\text{cm})$$

(b) If this capacitance is used to store the one bit of information at a power supply voltage of 3.2V, how many electrons are stored on the capacitance?

Use $C_{Omeas}=Q_G/V_G$ then the stored electron number is $Q_G/q = C_{Omeas}V_G/q = (5\times10^{-14}\text{F})(3.2\text{V})/(1.6\times10^{-19}\text{C/electron}) = 1.0\times10^6$ electrons.

A power supply voltage less than 3.2V is desirable because there would be fewer energetic electrons to surmount the 3.13eV SiO_2/Si barrier and to become trapped at defects in the oxide which would give rise to charges in the oxide, Q_{OT}. This trapping should be minimized during device operation, because oxide charge changes the gate threshold voltage, V_{GT}, affects the output characteristics, and reduces the device reliability. How do we find the barrier height? Use Table 413.1 and $\phi_{Barrier} = \chi_{Si} - \chi_{SiO2} = 4.029 - 0.9 = 3.13$eV.

(c) Where is the Fermi level measured relative to the intrinsic Fermi level, $E_F - E_I$, in eV? Give the result also in terms of the Fermi potential relative to the intrinsic Fermi potential, i.e. V_F =?.

$$N = n_i \exp[-(E_I - E_F)/kT)]$$
$$E_F - E_I = kT\times\log_e(N_{DD}/n_i) = (0.025\text{eV})\log_e(10^{17}/10^{10}) = 0.4027\text{eV}$$
$$V_F - V_I = -(E_F - E_I)/q = -0.4027\text{V}$$

(d) What is the threshold voltage at the onset of strong surface inversion?

$V_{GTH} = V_G(V_S = 2V_F) = V_{FB} + V_S + \varepsilon_S E_S/C_O$

$= V_{FB} + 2V_F - \sqrt{[4\varepsilon_S qN_{DD}|V_F|]}/C_O$ (411.9) Note: The sign of E_S for nMOSC is negative for inversion ($-V_G$) then

$\varepsilon_S E_S = \sqrt{[4\varepsilon_S qN_{DD}|V_F|]}$

$= \sqrt{[4\times8.854\times10^{-14}(F/cm)\times1.602\times10^{-19}\times1.0\times10^{17}cm^{-3}|V_F|]}$

$= 7.532\times10^{-8}\sqrt{|V_F|}C/cm^2.$ (P411.3A)

$V_{GTH} = 0+2(-0.402V)-7.53\times10^{-8}|-0.402V|^{1/2}(C/cm^2)/(5\times10^{-7}F/cm^2 = -1.1V$

The following capacitance calculations are much easier made if we calculate the total capacitance in fF unit or capacitance per micron^2 in pF/μm^2 rather rather than capacitance per cm^2 or m^2 in F/cm^2 and F/m^2 to avoid writing a large negative exponent of 10. The per micron unit is also the most practical since the dimension of current and future integrated circuits are of micron and submicron sizes.

(e) What is the capacitance of the MOSC at the very strong inversion condition?

$C_{g\infty} = C_oC_s/(C_o+C_s) = C_O/\{1 + C_O/\sqrt{[(\varepsilon_S qN_{DD}/2)/(2|V_F|+3kT/q)]}\}$ (411.31)

$C_s = [(\varepsilon_S qN_{DD}/2)/(2|V_F|+ 3kT/q)]^{1/2}$

$= [(7.532\times10^{-8}/\sqrt{8})\div(2\times0.4027+3\times0.025)^{1/2} = 9.73\times10^{-8}F/cm^2$

$= 9.73fF.$

$C_{g\infty} = (5\times10^{-7}F/cm^2)/(1+5\times10^{-7}/9.73\times10^{-8}) = 8.145\times10^{-8}F/cm^2 \Rightarrow 8.145fF$

(f) What is the capacitance of the MOSC at the threshold or onset of inversion?

$C_{gth} = C_O/(1 + C_O/\sqrt{[\varepsilon_S qN_{DD}/(4V_F)]})$

$= (5\times10^{-7}F/cm^2)\div\{1+(5\times10^{-7}F/cm^2)\div[(1.67\times10^{-14})/(4\times0.4027V)]^{1/2}$

$= 8.4\times10^{-8}F/cm^2 \Rightarrow 8.4fF$

(g) What is the d.c. voltage at the intrinsic surface condition?

$V_{GI} = V_G(V_S = V_F)$

$= V_{FB} + V_S + \varepsilon_S E_S/C_O$

$= V_{FB} + V_F - [2\varepsilon_S qN_{DD}|V_F|]^{1/2}/C_O$

$= 0 + (-0.4027V) - (7.532\times10^{-8}\sqrt{0.4027})/5\times10^{-7} = -0.63\ V$

(h) What is the capacitance of the MOSC at intrinsic surface?

$C_{gI} = C_O/(1 + C_O/\surd[\varepsilon_S q N_{DD}/(2V_F)])$
$C_{gI}A = 50fF/(1+50/14) = 11fF$
$C_{gI} = 11fF/10\mu m^2 = 1.1\times10^{-7}F/cm^2 = 1.1fF/\mu m^2$

(i) What is the d.c. voltage at flat band?

$V_{GFB} = V_G(V_S = 0)$
$= V_{FB} + 0 + \varepsilon_S E_S/C_O$
$= V_{FB} + 0 + 0 \quad = 0.$

The above null answer originates from $E_S = 0$ because the problem assumes zero oxide and interface trap densities and the gate material is not specified completely (no gate dopant concentration given) so we assume it to be identical the substrate except it is polycrystalline silicon, i.e. $N_{DD\text{-}gate} \equiv N_{DD\text{-}substrate} = N_{DD} = 1.0\times10^{17}cm^{-3}$ = given.

(j) What is the capacitance of the MOSC at flat band?

$C_{FB} = C_O/[\![1 + C_O/\{\varepsilon_S/\surd[(\varepsilon_S kT)/q^2(P_B + N_B)]\}]\!]$
$= (5\times10^{-7}F/cm^2)/[\![1+(5\times10^{-7}F/cm^2)/(11.8)(8.854\times10^{-14}F/cm)$
$\div\surd\{[(11.8)(8.854\times10^{-14}F/cm)(0.025V)]/[(1.6\times10^{-19}C)(10^{17}cm^{-3})]\}]\!]$
$= 3.14\times10^{-7}F/cm^2 = 3.14fF/\mu m^2 \Rightarrow 31.4fF.$

P411.4 A parallel voltage shift of the experimental HFCV curve of the 100A-oxide MOSC in problem P401.1 was observed when compared with the ideal theory. The observed voltage shift was –0.1V. (Since the problem did not state the gate material, we will assume identical gate and substrate, i.e. n-type with $N_{DD}=10^{17}cm^{-3}$ so the gate/semiconductor workfunction difference is zero, $\Phi_{MS}\equiv\Phi_{GS}=0$.)

(a) What is the density (Coulomb/cm^2) and sign of the oxide charge?

$C_O = \varepsilon_o/X_o = (3.9\times8.854\times10^{-14}F/cm)/(100\times10^{-8}cm) = 3.45\times10^{-7}F/cm^2$
$Q_{OT} = -C_O\Delta V = -(3.45\times10^{-7}F/cm)(-0.1V) = 3.45\times10^{-8}$ Coulomb/cm^2.

The sign of the oxide charge is <u>positive</u> for a negative gate voltage shift.

(b) What is the oxide trap density if each trap is singly charged?

$N_{OT} = |Q_{OT}/\mathbf{z}q|$ where **z**=1 for singly charged oxide traps.

$N_{OT} = |(3.45\times10^{-8}\text{Coul/cm}^2)/(1.6\times10^{-19}\text{Coul/1-oxide-trap})$

$= 2.1\times10^{11}$ oxide-trap/cm^2.

P411.11 The flat-band capacitance and the Debye length concept described in section 411(E) for the p-Si MOSC are actually more general and apply to any region of a semiconductor. Demonstrate their validity in a p-type semiconductor region at equilibrium, labeled x=0, where the dopant impurity concentration varies with position as $N_{AA}(x) = N_{AA}(0) + ax$, where the the dopant impurity concentration gradient, a, is a constant. What is the restriction on 'a' in order that the validity is accurate? (Hint: The concentration gradient 'a' is much smaller than a characteristic-concentration divided by a characteristic-length).

This can be obtained by asking the question of how small can 'a' be while the region x=0 is still nearly electrically neutral, i.e. quasi-neutral is still a good approximation.

To solve this, we expand the Boltzmann expression of the electron and hole concentration by power series at x=0 and retain only the first term and using this in the Poisson equation

$$-\varepsilon d^2V/dx^2 = q[P - N - N_{AA}(0) - ax]$$
$$\simeq q[2n_i qV/kT - N_{AA}(0) - ax]$$

Then the quasi-neutrality condition is approached at x=0 if $a < N_{AA}(0)/L_D$ where L_D is the Debye length at x=0. This can also be transformed to give a criteria of small electric field: $|E(x=0)| << kT/qL_D$.

P411.12 It is evident from subsection 411(E) that the Debye length is a characteristic parameter of a semiconductor. Some semiconductor mathematicians and device physicists have used the intrinsic Debye length which is about 29μm in Si at 300K while the extrinsic Debye length is 410A at $N_{AA} = 10^{16}$ cm^{-3} computed from (411.17). It is evident that the intrinsic Debye length is not very meaningful. What is the intrinsic Debye length of Si, Ge, GaAs, and GaP at 300K? Use the data of n_i from chapter 2. What is the Debye length (or extrinsic Debye length) of these semiconductors doped to 1.0×10^{16} cm^{-3} of donor impurities? Explain the usefulness of the extrinsic Debye length and the uselessness of the intrinsic Debye length.

Using Fig.241.1 for n_i, the intrinsic Debye length is computed from

$$L_{DBi} = \sqrt{(\varepsilon_S kT/2q^2 n_i)}.$$

(a) Si: $\varepsilon_S=11.8\varepsilon_0$, $n_i(300K)\approx1\times10^{10}cm^{-3}$ $L_{DBi}(Si) = 2.9\times10^{-3}cm = 29\ \mu m$
(b) Ge: $\varepsilon_S=16.0\varepsilon_0$, $n_i(300K)\approx2\times10^{13}cm^{-3}$ $L_{DBi}(Ge) = 7.4\times10^{-5}cm = 0.74\mu m$
(c) GaAs: $\varepsilon_S=13.1\varepsilon_0$, $n_i(300K)\approx2\times10^{6}\ cm^{-3}$ $L_{DBi}(GaAs) = 0.21cm$
(d) GaP: $\varepsilon_S=11.1\varepsilon_0$, $n_i(300K)\approx3\ cm^{-3}$ $L_{DBi}(GaP) = 160\ cm$

The extrinsic Debye length is calculated from (411.26A)

$L_{Bi} = \sqrt{[\varepsilon_S kT/q^2(P_B+N_B)]}$. Since
$N_{DD} = 10^{16}cm^{-3} > 10n_i$, $N_B\approx N_{DD}$ and
$P_B = n_i^2/N_{DD} << N_B$, therefore
$L_{DBi} = \sqrt{(\varepsilon_S kT/q^2 N_{DD})}$:

(e) Si: $\varepsilon_S=11.8\varepsilon_0$ $L_{DB}(Si) = 4.04\times10^{-6}cm = 404$ Angstroms
(f) Ge: $\varepsilon S=16.0\varepsilon_0$ $L_{DB}(Ge) = 470$ Angstroms
(g) GaAs: $\varepsilon S=13.1\varepsilon_0$ $L_{DB}(GaAs) = 426$ Angstroms
(h) GaP: $\varepsilon S=11.1\varepsilon_0$ $L_{DB}(GaP) = 392$ Angstroms

The extrinsic Debye length is an important characteristic length of an extrinsic semiconductor because it indicates how far a small-signal charge can propagate, i.e. it is the decay length of the amplitude or number in a pulse of electron from its source position, $n(x) = n(0)\exp(-x/L_{DB})$. For another example, the capacitance at flatband is $C_{sfb} = \varepsilon_S/L_{DB}$. In comparison, the intrinsic Debye length does not appear in any semiconductor device characteristics and even the i-region (i=intrinsic) high voltage p-i-n rectifiers has a dopant impurity concentration a few orders of magnitude higher than n_i. One of the device with a nearly intrinsic region is the alpha and beta particle semiconductor detector used in nuclear physics.

Chapter 5
P/N AND OTHER JUNCTION DIODES

OBJECTIVES

* Additional Device Physics (over MOSC)

- Minority carrier diffusion-generation-recombination currents.
- Current multiplication, divergence at breakdown voltage.
- Minority and majority carrier trapping transients.
- Thermionic current (a barrier surmounting injection).
- Tunneling current (negative conductance).
- Nonequilibrium energy band diagram.
- Heterojunction energy band and energy gap alignment.

* Device Analysis Technique Using Differential Equation

⇒ Application of energy band diagram.

- Solution of diffusion differential equation.
- Differential charge analysis of capacitance and conductance.
- Small-signal sinusoidal steady-state admittance versus signal frequency.
- Large-signal charge-control transient.
- Trapping transient in reverse differential capacitance and d.c. current.

* Device Characteristics

- D. C. current-voltage, $I(V)$.
- Capacitance- and conductance-voltage, $C(V)$, $G(V)$.
- Capacitance- and conductance-frequency, $C(\omega)$, $G(\omega)$.
- Switching transients.
- Reverse-bias trapping transients, $C_{rev}(t)$, $G_{rev}(t)$, $I_{REV}(t)$.

* Circuit Models, Symbols

- D. C. model (symbol).
- Small-signal model.

* Additional Device Process Chemistry and Design (over MOSC)

- Solid-state diffusion, diffusivity physics.
- Design analysis for thin p-layer of p/n junction.
- Oxide masking against diffusion (see chapter 6 on MOST).

P512.4 First generation (1960) ultrahigh speed switching Si diodes and transistors such as the 1N914 and 2N706 used in the first supercomputer, the Control Data Corporation model 6600 attained their speed by doping with the gold recombination center to a concentration of $10^{16}cm^{-3}$ because gold has two very efficient recombination levels (see Fig.381.2) and gold can be diffused into Si quickly at low diffusion temperatures due to its high diffusivity in Si (see Fig.512.2). Suppose that this high-temperature gold diffusivity in Si can be extrapolated to room temperature (300K) using the Arrhenius relationship. (a) What is the gold diffusivity in Si at 300K and (b) what is the time to failure (TTF) because the substitutional gold atoms jump (a diffusion step) to adjacent interstitial sites 5A away and assume that the interstitial gold is not an electronic trap or a recombination center?

There are two ways to estimate the gold diffusivity at 300K using the high temperature experimental data shown in Fig.512.3: graphically extend the straight line to 300K or 1000/T=1000/300=3.333 of the upper linear x-axis since the figure plots $\sqrt{D}$ vs 1/T which is a straight line because the diffusivity follows the Arrhenius relationship given by (512.1), $D = D_0\exp(-E_A/k_BT)$, therefore $\sqrt{D} = \sqrt{D_o}\exp(-E_A/2k_BT)$ which is a straight line when plotted on a semilog paper of $y=\log_{10}\sqrt{D}$ vs $x=1000/T$ using $\log_{10}\sqrt{D} = \log_{10}\sqrt{D_0} + (\log_{10}e)\times(-E_A/2k_BT)$. The second method is to pick two points on the experimental line in Fig.512.3 and solve for the two parameters D_0 and E_A in the diffusivity equation, (512.1), then use the equation and the two parameters' value to compute the diffusivity at 300K.

In either way, the diffusivity of gold at 300K is about $D_{Au}(300K) \sim (2\times10^{-9}cm)^2$/hour. (Using a two point fit, we got $\sqrt{D} \sim 66000\times\exp(-6600/T)$ $\mu m/\sqrt{hr}$.) If we estimate the distance traveled by a gold atom from $\Delta x \sim (Dt)^{1/2}$ then the time required for a gold atom to jump a distance of $\Delta x \sim 5A$ or $5\times10^{-8}cm$ at 300K is approximately $t \sim (\Delta x)^2/D = (5\times10^{-8}cm)^2/(2\times10^{-9}cm)^2$ hours = 625 hours.

This is obviously too short even if the numerical estimate of D is not accurate or off by 10 or even 100 since the gold-doped Si bipolar transistors and diodes made by this author at the Fairchild Semiconductor Laboratory 35 years ago (1961) have not shown any significant change in their characteristics. There are two possible errors in this problem and its solution just presented. (1) The transistor's and diode's characteristics do not change by just one gold atom moving 5A. The switching time would increase perceptibly only when the gold concentrations in the minority carrier recombination volume (about $1\mu m\times10\mu m\times10\mu m$) is decreased by 10% via diffusion to the silicon chip's surface which is a sink of gold and which is located at about 5μm away from the recombination volume. This would increase Δx from 5A to 5×10^4A or increase the TTF by $(10^4)^2$ to 6.25×10^{10} hours or about 7×10^6 years which would make the migration of gold irrelevant or negligible as far as 1960-transistor failure is concerned. (2) However, experiments have not shown gold movement at much smaller dimensions than the 5μm used in (1), such as 100A. Thus, another potential source of error is the inapplicability of the Arrhenius relationship over such a large temperature range. At 300K the process that dominates gold migration in silicon may be different from that in the 800C-1200C range where the gold diffusivity data in Fig.512.3 were obtained experimentally. For example, substitution gold migration from one substitution site to another substitution site could involve three steps, a jump from the substitution site to the adjacent interstitial site by breaking the Au-Si bond, diffusion from the interstitial to adjacent interstitial site, and then a jump into an adjacent silicon vacancy. The bond energy in the first step is temperature dependent, not unlike the Si-Si bond energy we estimated by (512.2) in the text to create Si vacancy which drops from 1.21eV at room temperature (300K) to about half or 0.5eV at 1300K. This would make the D vs 1000/T curve concave downwards below the high-temperature straight line shown in Fig.512.3 as the temperature is lowered. There could also be another migration pathway in which the substitution gold atom would jump into an adjacent silicon vacancy which is 2.35A away on the nearest lattice site. This would then not involve an interstitial diffusion step as that in the three-step pathway just described. But still the bond breaking energy to make the jump would dominate which would give a strongly temperature dependent thermal activation energy that would prevent a straight line extrapolation of the high temperature gold diffusivity to 300K. The student may extend this solution to take into account of the increasing bond energy with lowering temperature using E_G vs T given by (512.2) as the scale.

P513.1 A p+/n junction is to be fabricated by a two-step diffusion procedure involving an initial pre-deposition step of an amount, Q(atoms/cm^2), of boron impurity atoms and a final drive-in step at a higher temperature to drive in the deposited impurity atoms by diffusion to a required junction depth, x_j (μm), and to achieve a required surface concentration of C_o (atom/cm^3). The substrate n-type donor dopant impurity concentration is C_B (atom/cm^3). Given: $C_0=1.0\times10^{19}$cm^{-3}, $C_B=1.0\times10^{15}$cm^{-3}, $x_j=1.0$μm, and $T_d=1000$°C. (a) What is the diffusion time required? (b) What is the predeposited Q?

The concentration profile of the deposited-diffused boron impurity follows the Gaussian function given by (513.3) from which (a) the diffusion time, t_d, and (b) the predeposited Q can be computed. To start, we need the diffusivity of boron in silicon at 1000°C which is given in Fig.512.2 which we will assume to be valid or to give a good estimate at all boron and n-type donor dopant impurity concentrations. Thus, $\sqrt{D_{boron}} \simeq$ 0.13 μm/√hour at T_d=1000°C. From (5.13.3),

(a) $C(x_j) = C_B = 1.0\times10^{15}cm^{-3} = C_0 \exp(-x_j^2/4D_{boron}t_d)$

$= 1.0\times10^{19}cm^{-3}\times\exp[-1.0^2\mu m^2/(4\times0.13^2\times t_d)]$

$\therefore\ t_d = (1.0/4\times0.13)^2/\log_e(10^{19}/10^{15}) = 1.606$ hours.

(b) $C_o = Q/\sqrt{(\pi Dt)}$.

$Q = C_o\sqrt{(\pi Dt)} = 10^{19}\times0.13\times10^{-4}\sqrt{\pi\times1.606} = 2.9\kappa10^{14}cm^{-2}$.

P522.1 A diffused Si p+/n junction is to be approximated or modeled by an abrupt Si p+/n junction. The abrupt model is defined by N_{AA} = 1.0 x 10^{19} cm^{-3} = constant from x=0 to x_j and N_{DD} = 1.0x10^{15} cm^{-3} = constant from x_j to the back surface and x_j = 1.0 μm. What are the following properties at equilibrium, $V_{Applied}$ = 0, at T=296.6K at which $n_i \simeq 1.0\times10^{10}$ cm^{-3} and kT/q ≃ (should read 0.025V).

(a) The potential barrier height, V_B, (which is V_{bi} at V=0V) in (mV).

$$V_{bi} = (kT/q)\times\log_e(N_{AA}N_{DD}/n_i^2) \qquad (522.7A)$$

$$= (25mV)\times\log_e[10^{19}10^{15}/(10^{10})^2] = 805.9\ mV.$$

(b) The total, space charge layer thickness, x_{pn} (μm).

$$x_{pn} = \sqrt{[(2\varepsilon_S V_{bi})/(qN_M)]}\ \text{where} \qquad (523.19)$$

$$N_M = (N_{AA}N_{DD})/(N_{AA}+N_{DD}) = (10^{19}cm^{-3}10^{15}cm^{-3})/(10^{15}cm^{-3}+10^{19}cm^{-3})$$

$$= 1.00\times10^{15}\ cm^{-3},\ \text{therefore,}$$

$$x_{pn} = \sqrt{[(2\times11.7\times8.854\times10^{-14}F/cm\times0.806V)/(1.6\times10^{-19}C\times1.00\times10^{15}cm^{-3})]}$$

$$= 1.03\times10^{-4}\ cm = 1.03\ \mu m.$$

(c) The space charge layer thicknesses on the p-side and n-side, x_p and x_n in (μm).

$$x_p = (x_{pn}N_{DD})/(N_{AA} + N_{DD}) \qquad (523.23C)$$

$$= (1.03\mu m)\times(10^{15}cm^{-3})/(10^{15}cm^{-3}+10^{19}cm^{-3}) = 1.03\times10^{-4}\mu m.$$

$$x_n = (x_{pn}N_{AA})/(N_{AA} + N_{DD}) \qquad (523.23C)$$

$$= (1.03\mu m)\times(10^{19}cm^{-3})/(10^{15}cm^{-3}+10^{19}cm^{-3}) = 1.03\mu m$$

(d) The maximum electric field, E_{MAX} (V/cm).

$$E_{MAX} = \surd(2qV_{bi}N_M/\varepsilon_S) \quad (523.21)$$

$$E_{MAX} = 2V_{bi}/x_{pn} \quad (523.22)$$

$$= 2\times0.8059V/1.03\times10^{-4}cm = 1.56\times10^4V/cm.$$

P522.2 Derive the V_{bi} formulae using $J_P = 0$ and the textbook procedure using $J_N = 0$.

At equilibrium, the net electron current and the net hole currents are each equal to zero. We used J_N=0 to derive the formula for V_{bi} in the textbook. Now, let us use the hole current given by (320.11) which is set to zero as follows.

$$0 = J_{Px} = q\mu_pPE_x - qD_p(dP/dx) \quad (320.11)$$

$$\therefore qD_p(dP/dx) = q\mu_pP(-dV_I/dx)$$

$$(D_p/\mu_p)(dP/P) = -dV_I$$

Use the Einstein relation D_p/μ_p=kT/q and integrate the last equation above from location x_1 to x_2, we have

$$\int(kT/q)(dP/P) = \int-dV_I$$

$$(kT/q)\ \log_e[P(x_2)/P(x_1)] = -[V_I(x_2) - V_I(x_1)].$$

We make use of the logarithmic relationship $-\log_e(A/B) = +\log_e(B/A)$, then

$$(kT/q)\times\log_e[P(x_1)/P(x_2)] = V_I(x_2) - V_I(x_1).$$

Let $x_2 = +T_n$ (n-Si) where $P(x_2=T_n) = N_N/n_i{}^2$.

Let $x_1 = -T_p$ (p-Si) where $P(x_1=-T_p) = P_P$.

Then, $(kT/q)\times\log_e[P_PN_N/n_i{}^2] = V_I(x_2=T_n) - V_I(x_1=-T_p) = V_{bi}$

This is exactly the expression for the built-in potential, (522.2) obtained using J_N=0 as the starting point, showing anticipated consistency. **Q.E.D.**

P522.5 Diodes and transistors are no longer useful when the temperature is so high that the p/n junction built-in potential barrier height is less than about 4kT. Calculate this temperature for an Si p+/n abrupt junction with $N_{AA}=10^{18}cm^{-3}$ and $N_{DD}=10^{16}cm^{-3}$.

From (522.2A), $n_i^2 = N_{AA}N_{DD}\exp(-qV_{bi}/kT) = 10^{18}\times10^{16}\exp(-4)=(1.353\times10^{16})^2$.
From the Si n_i-T curve in Fig.241.1(a), we get $T = 380C$ at $n_i \sim 1.353\times10^{16}cm^{-3}$.

P534.1 A forward voltage of V is applied to a thin-base p+/n/m diode whose n-base is 1 μm thick, i.e. T_n = 1μm. And $T_n << L_p = \sqrt{(D_p\tau_p)}$ = 100μm. The n/m back surface contact is perfect (zero resistance). Sketch P(x) and $J_P(x)$ in the n-type quasi-neutral base layer without solving the hole diffusion equation. Label the concentrations and current densities at the boundary x_n and T_n.

The hole concentration and hole current density in the n-type quasi-neutral base layer can be sketched without solving the hole diffusion equation by looking at the general solution and considering the boundary conditions.
General solution of the hole diffusion equation:

$$P(x) - P_N = A\exp(x/L_P) + B\exp(-x/L_P). \qquad (532.7)$$

Boundary conditions:

$$P(x=x_n) = P_N\exp(qV/kT) \qquad (532.9)$$

$$P(x=T_n) = P_N$$

If $T_n << L_p$, then $x << L_p$ and $\exp(\pm x/L_p) \approx 1 \pm x/L_p$.
Therefore, the hole concentration depends linearly on x between $x=x_n$ and $x=T_p$ and must satisfy the boundary conditions above.

Thus, we can sketch P(x). Using the boundary conditions, we can write down the equation for P(x):

$$P(x) = P_N[\exp(qV/kT) - 1][1 - (x-x_n)/(T_n-x_n)] + P_N.$$

Note that this equation is linearly decreasing in x and satisfies the B.C. given above.

The hole current density is proportional to the slope of the hole concentration since $J_P(x) = -qD_p(dP/dx)$. Therefore, the hole current density is a constant because the hole concentration decreases linearly with x. Thus, we can also sketch $J_P(x)$ = constant. The value of the hole current density is given by:

$$J_P(x) = -qD_p(dP/dx) = [(qD_pP_N)/(T_n-x_n)][\exp(qV/kT) - 1]$$

P537.1 Continue on to solve for the remaining characteristics of the p+/n diode given in problems P522.1, P522.5 and P523.1. Assume $\tau_p=\tau_n$=1.0μs, $n_i=1.0\times10^{10}$cm^{-3}, and kT/q=0.025V (T=296.6K). Use the appropriate majority and minority carrier mobilities from figures or formulae in chapter 3.

(a) Calculate J_1 of the Shockley and J_2 of the SNS current components in A/cm^2.

Shockley Ideal Diode Current:

Use $J_1 = J_{p0} + J_{n0} = qD_pP_N/L_p + qD_nN_P/L_n$ (532.15)

n-side:

$\mu_p(N_{DD}=1.0\times10^{15}\ cm^{-3}) = 460\ cm^2/V\text{-}s$ (from Fig. 313.5)

$D_p = \mu_p(kT/q) = (460\ cm^2/V\text{-}s)(0.025V) = 11.5\ cm^2/s$

$L_p = \surd(D_p\tau_p) = \surd((11.5\ cm^2/s)(1\times10^{-6}\ s)) = 3.39\times10^{-3}\ cm$

Use $N_NP_N = n_i^2$ and $N_N \approx N_{DD} = 10^{15}\ cm^{-3}$; then

$P_N = n_i^2/N_N = (10^{10})^2/10^{15} = 10^5\ cm^{-3}$

$J_{p0} = qD_pP_N/L_p = (1.6\times10^{-19}\ C)(11.5\ cm^2/s)(10^5\ cm^{-3})/(3.39\times10^{-3}\ cm)$

$= 5.426\times10^{-11}\ A/cm^2$

p-side:

$\mu_n(N_{AA}=1.0\times10^{19}\ cm^{-3}) = 130\ cm^2/V\text{-}s$ (from Fig. 313.5)

$D_n = \mu_n(kT/q) = (130\ cm^2/V\text{-}s)(0.025V) = 3.25cm^2/s$

$L_n = \surd(D_n\tau_n) = \surd((3.25\ cm^2/s)(1\times10^{-6}\ s)) = 1.80\times10^{-3}cm$

Use $N_PP_P = n_i^2$ and $P_P \approx N_{AA} = 10^{19}\ cm^{-3}$, then

$N_P = n_i^2/P_P = (10^{10})^2/10^{19} = 10^1cm^{-3}$.

$J_{n0} = qD_nN_P/L_n = (1.6\times10^{-19}C)(3.25cm^2/s)(10^1cm^{-3})/(1.80\times10^{-3}cm)$

$= 2.884\times10^{-15}\ A/cm^2$

$J_1+J_{p0}+J_{n0} = 5.426\times10^{-11}A/cm^2 + 2.884\times10^{-15}A/cm^2 \approx 5.426\times10^{-11}A/cm^2$.

SNS Recombination Current in the Junction Space Charge Layer:

Use $J_2 = [qn_i/(\tau_{n0}+\tau_{p0})]\times(x_p+x_n)$ (535.3)

From the solution to P522.1,

$x_p = 1.03\times10^{-4}\mu m$ and $x_n = 1.03\mu m$ where $1\mu m = 10^{-4}cm$, therefore,

$J_2 = (1.6\times10^{-19}C)\times(10^{10}cm^{-3})\times(1.03\times10^{-8}cm+1.03\times10^{-4}cm)$

$\div (1.0\times10^{-6}s+1.0\times10^{-6}s)$

$= 8.240\times10^{-8}\ A/cm^2$

(b) Look up the breakdown voltage of the junction from Fig. 536.2.

From Fig. 536.2 using $N_{ION} = N_{DD} = 10^{15}cm^{-3}$ for the p+/n junction diode, the breakdown voltage is about 300V.

(c) Compute the threshold injection voltage as defined by $J_{SHOCKLEY}=J_{SNS}$ discussed at the end of section 535.

From $V_{IT} = 2(kT/q)\log_e(J_2/J_1)$ (535.7)

$V_{IT} = 2(0.025V)\log_e[(8.240\times10^{-8}A/cm^2)/(5.426\times10^{-11}A/cm^2)] = 0.366V$

(d) Sketch the I-V curves in the three I and V scales similar to those used in Fig. 500.1. Select the appropriate range of I and V scale for each of the three figures in order to show the details of the variations. Read hints in text to make the back of the envelope sketches while still get very accurate plots. Computer generated point-by-point plots are not asked for by the problem and will not receive credit.

P541.4 What is the physical significance when $V=V_{bi}$ which makes $x_{pn}=0$ and $C_d=\infty$?

When $x_{pn}=0$, the electron and hole charges stored in the layer x_{pn} is zero and they would stay zero when the applied voltage V is further increased to larger values than V_{bi}. The depletion capacitance is meaningless since the depletion layer no longer exists. At some voltages several kT/q below V_{bi}, the depleted space-charge layer concept no longer applies because of the presence of a larger concentration of both electrons and holes in the space-charge layer injected by the applied forward bias voltage. The appropriate formulae to calculate the differential capacitance is dQ/dV where Q can be either the integrated electron or hole concentration over the entire volume of the p/n junction diode from the ohmic contact to the p-surface to the ohmic contact to the n-surface.

P542.3 A silicon p/n junction diode is used as a voltage tunable capacitor to automatically scan the AM radio stations in the 550kHz to 1650kHz band. What is the applied d.c. voltage range needed to tune this 3:1 frequency range assuming that you have a 12 Volt battery? (A tuned parallel L-C circuit has a resonance or peak impedance frequency of $f = 1/2\pi\sqrt{LC}$.)

The 3:1 frequency range requires a $(3:1)^2 = (9:1)$ capacitance variation because the frequency is proportional to the $\sqrt{C}$. If the p/n junction diode has an abrupt constant impurity profile, i.e. N_{DD} and N_{AA} are spatially constant, then $C \propto (V_{bi} + V)^{-1/2}$. Thus, the voltage range must cover $(V_{bi} + 12)/(V_{bi} + 0) = 9^2 = 81$ or the build-in voltage must be less than $12/80 = 0.15V$. This is too small to be fabricated therefore the applied voltage must be larger, about 5 times larger or 60V to give a easily fabricateable V_{bi} of 0.75V. One alternative solution was to make the capacitance voltage variation much larger than the square-root dependence by impurity profile engineering (the hyper-abrupt profile). A better alternative was to divide the 3:1 band into three bands and to switch in a fixed parallel capacitor at the two boundaries of the three bands, or to use two or three variable capacitors to cover this 3:1 AM radio band so that each capacitor covers a smaller range.

The voltage-variable p/n junction capacitor for station tuning was much more practical in the higher frequency bands using a small d.c. voltage and voltage swing. For examples, in the shortwave and FM broadcast receivers, the bandwidth/center-frequency ratio is very small compared with the 3:1 ratio (550-1650kHz) of the AM broadcast band even the bandwidth is similar to or much wider than the AM radio's 1100kHz. FM radio's 20MHz bandwidth (88-108MHz) is only 20/98≃20%. The 15MHz/20-meter shortwave broadcast band's 500kHz bandwidth (15-15.5MHz) is only 0.5/15≃3.3%. This analogy radio electrical tuning scheme was a much improvement over the mechanical variable capacitor but it was soon replaced by precision quartz-crystal oscillator and inexpensive digital synthesizer circuit to generate extremely precise local oscillator frequencies for superheterodyn mixing and detection, which precision cannot be attained by stabilizing and digitizing the tuning voltage applied to p/n junction capacitor.

P542.5 A long (or thick) silicon p/n junction diode, such as the numerical example given in the text, is made on a silicon wafer of 1mm thick so that the series bulk resistance cannot be neglected. Calculate the cutoff frequency due to the series resistance when the diode is reverse biased, zero, and forward biased at the three voltages in Table 542.1. How do these cut-off frequencies compare with the diffusion-recombination delay (roughly the lifetime or G/C time constant and the charging time of the space-charge layer)? Under what application conditions is the series resistance important?

The p+/n diode in Table 542.1 has $N_{DD}=10^{14}cm^{-3}$ and $D_n=25cm^2/s$ at T=297K or kT=0.0255eV. The majority carrier (electrons) mobility in the L=1mm=0.1cm thick n-Si is then $\mu_n = D_n/(kT/q) = 25/0.0255 = 1000\ cm^2/V\text{-}s$. The majority carrier (electron) concentration is approximately $N \simeq N_{DD} = 10^{14}cm^{-3}$ since most of the donor impurity atoms at this temperature are ionized or released its extra (5-th) valence electron to the silicon conduction band to become a conduction electron. Thus, the resistivity of the 1mm thick n-Si is $\rho = (q\mu_n N)^{-1} = (1.602\times10^{-19}\times1000\times10^{14})^{-1}$ ohm-cm = 62.42 ohm-cm.

The RC time constants at the four voltages are then computed from

$$RC = (\rho L/A)\times(C_n+C_p+C_{pn})\times A = \rho L(C_n+C_p+C_{pn}) = (6.242\text{ohm-cm}^2)\times(C_n+C_p+C_{pn})$$
$$= (6.242\text{ohm-cm}^2)\times(2.40\times10^3\text{pF/cm}^2) = 15\text{ns at } V=-0.1V \quad C/G = 24\text{ms}$$

$= \quad 3.73\times10^{3} \quad = \quad 24\text{ns at V= 0.0V} \quad C/G = 858\text{ms}$

$= \quad 6.16\times10^{6} \quad = 38.4\mu\text{s at V=+0.5V} \quad C/G = \quad 1\mu\text{s}$

It is evident that the series resistance is important only at large forward bias such as the 0.5V example given above. At zero and reverse biases, the diode loss conductance is negligible compared with the series resistance which would limit the diode's frequency response. The two numerical examples show frequency response drop off begins at $1/2\pi RC$ = 10.6MHz at V=–0.1V and 6.63MHz at V=0.0V.

P553.4 Obtain the switching transient solutions for a thin-base diode with the p+/n/m structure where n is the thin base of geometrical thickness X_B and $X_B << L_B$. n/m is a perfect ohmic contact of infinite recombination rate. Use the charge control model and assume minority carrier diffusion to be dominant. The mathematics is very simple. Students are to give simple physics-based explanation.

The forward current is

$$J_F = qP_N[\exp(qV_{PN}/kT) - 1]D_p/X_B.$$

The hole charge stored in the thin base is

$$Q_P = qP_N[\exp(qV_{PN}/kT) - 1]X_B/2.$$

Thus, the charge storage time is

$$t_S = Q_P/J_F = X_B^2/2D_p.$$ (Recombination loss at n/m contact dominant.)

P553.5 Obtain the switching transient solutions for a thin-base diode with the p+/n/n+ structure where n is the thin base of geometrical thickness X_B and $X_B << L_B$ but recombination in the thin base-layer must not be omitted. n/n+ is a perfect donor contact of zero minority carrier (electron) current. Use the charge control model and assume minority carrier diffusion to be dominant. The mathematics is very simple. Students are to give simple physics-based explanation.

The forward current is

$$J_F = qP_N[\exp(qV_{PN}/kT) - 1]X_B/\tau_B.$$

The hole charge stored in the thin base is

$$Q_P = qP_N[\exp(qV_{PN}/kT) - 1]X_B.$$

Thus, the charge storage time is

$$t_S = Q_P/J_F = \tau_B.$$

(Recombination loss in thin base only because there is no recombination at n/n+.)

Chapter 6
METAL-OXIDE-SEMICONDUCTOR AND OTHER FIELD-EFFECT TRANSISTORS

OBJECTIVES

- Review and selection of concise device acronym.
- Definition of device structure terms.
- Quality discussion of device characteristics and physical origin
- Delineation of four basic device characteristics
- Sample fabrication steps
- Elementary analysis based on conductivity modulation model
- Charge-control analysis based on conductivity modulation model
- High-frequency response
  - Transconductance cutoff frequency
  - Gain-bandwidth product
- Switching properties
  - Intrinsic delay
  - Power-delay product
- Extrinsic delay - charging and discharging capacitance loads
- Circuit applications of MOSFET
  - Circuit symbol evolution and choice
  - DRAM cell
  - MOS inverter circuits
  - SRAM
  - Nonvolatile random access MOS memories

Note: Many new problems were added during fall-1993 when Sah taught the course again which did not appear in chapter 6's problem section 699 of the 1991 edition of FSSE. These were assigned to the students and solutions were given out which are collected herein.

P610.1 Draw the three-dimensional figure of a pMOST with a doped channel similar to that of Fig. 610.1 (which is an induced channel nMOST) using Fig. 623.1 as a guide. Label all parts of the pMOST structure and describe each part in analogous to the description given in Fig. 610.1.

Description of the Doped-Channel pMOST (following Fig. 610.1)

Coordinate System Same as Fig. 610.1.

y-axis length direction (longitudinal direction of channel current flow).
x-axis depth direction into the silicon body (direction transverse to current).
x-axis in the oxide film (oxide electric field controls the channel current).
z-axis width direction (gate width and channel width).

Body of Semiconductor (denoted by symbol X.)

- n-type Si bulk if there is a n/n+ LO-HI junction well. Body and Si bulk are at the same potential.
- n-type Si well if there is a n-well/p-bulk isolation junction well. Body (still designated as well) may not be at the same potential as the Si bulk.
- X is the contact to the n-body or the n-well. For the n-well/p-bulk case, the silicon body or bulk must use a different symbol than X which is reserved for the well or the body of the channel. No universal standard is employed. Thus, SUBSTRATE will be used with the symbol E for the node in some BiMOS and BiCMOS applications as the emitter of a vertical bipolar transistor or BE stands for Bottom Emitter and V_{BE} would be the voltage applied to E relative to B or X or the well, i.e. $V_{BE}=V_{XE}$.

Well

Either a n-well/n+bulk LO-HI junction well to have a low conductivity body on a n+ bulk Si-substrate, or a n-well/p-bulk (sometimes n-well/p-layer/p+bulk) n/p rectifying junction well to electrically isolate the n-body (the n-well) from the p-bulk. See above on labeling the p-substrate or p-bulk silicon wafer.

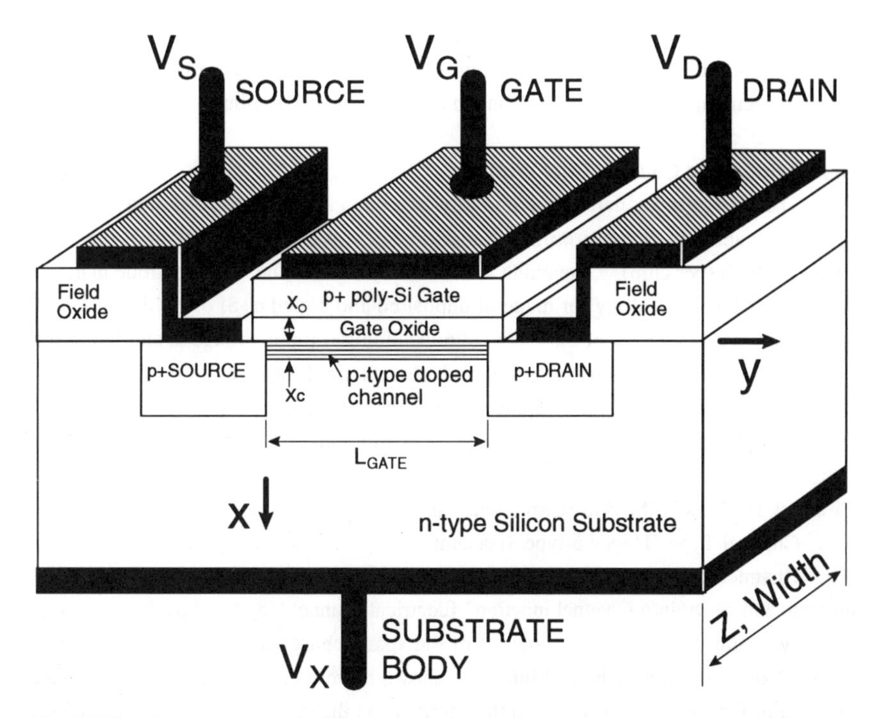

Fig.P610.1 The three-dimensional view of a doped p-channel Si MOST for defining the nomenclatures, physical dimensions and electrical parameters described in the text. The dark 45-degree shaded region are doped low resistivity poly-Si, silicide or metal layers.

Source and Drain

Two p+Si regions obtained by diffusing or implanting boron through two oxide windows (which define the area of the source and drain) into the n-Si bulk or n-Si well. The + sign in p+ means very high boron-acceptor concentration. The surface of the p+source and p+drain are contacted by a metal or high-conductivity polycrystalline silicon or silicide film which extends over the thick oxide (Field Oxide) to make connection with the adjacent transistors, or to enable contact with wires and electrical-connection pins of the transistor package (known as Contact Pad).

Channel

A thin doped p-type silicon layer underneath the metal-gated oxide. This doped p-type layer is the origin of the term 'doped channel'. It connects the p+source and p+drain electrically, and forms a p/n junction with the n-well or n-bulk.

Gate Oxide, Gate Contact, Gate Width

Thermally grown, defect-free/impurity-free thin gate oxide of 50A=5nm for 1996- and next generation of multi-million transistor Si chips to 2000A=200nm of the 1970 single MOSTs (per chip). Gate oxide is covered by a gate conductor electrode made of Al-metal (1970 technology) or made of deposited and doped n+Si or p+Si film (1996 technology). A 5V (3.3V or 2.5V for 1996 technology) input voltage is applied to the gate electrode (Gate Contact) creating high electric field through the thin oxide film. Gate Width is the mask or lithographic width of the gate electrode with symbol Z during 1960's and W for current VLSI-ULSI chips. Z is still used for power MOSTs.

Channel Type, Length, Width, and Thickness

Channel Type: Doped p-type Si channel.

Channel Length: Lithographic or mask length L between the p+Source/p-Channel junction and p+Drain/p-Channel junction. Electrical channel length is $L_E = L - y_{SC} - y_{DC}$ where y_{SC} and y_{DC} are the thicknesses of the space-charge layer of the p+Source/p-Channel and p+Drain/p-Channel junctions shown more quantitatively in the geometry diagram of Fig.685.2 in FSSE and in the energy band diagram of Figs.B1.1 and B1.2 on p.386-387 of FSSE-SG.

Channel Width: Electrical width of the conduction channel, W_E, may be smaller, equal or larger than a well-defined (i.e. constant oxide thickness without bird's beak) geometrical gate conductor width W_G due to fringe oxide electric field at the two edges in the z-direction (or width-direction) of the gate.

Channel Thickness: Electrical thickness of the conduction channel in the x-direction. In this doped channel case, it is thinner than the doped p-channel thickness because of the p-channel/n-body junction space-charge layer extension into the doped p-channel layer.

Field Oxide, Pad Oxide

Normally 5000A thick oxide underneath the metal (aluminum) lines that interconnect adjacent MOST's or underneath the contact metal pads that connect a wire to the pin of the package of the integrated circuit chip. Thick oxide reduces load capacitance, electrical field (high field causes transistor failure) and mechanical damage during wire bonding.

P630.1 Give the ten or more steps necessary to make the doped channel pMOST similar to those steps described in Fig. 630.1 for the induced channel nMOST.

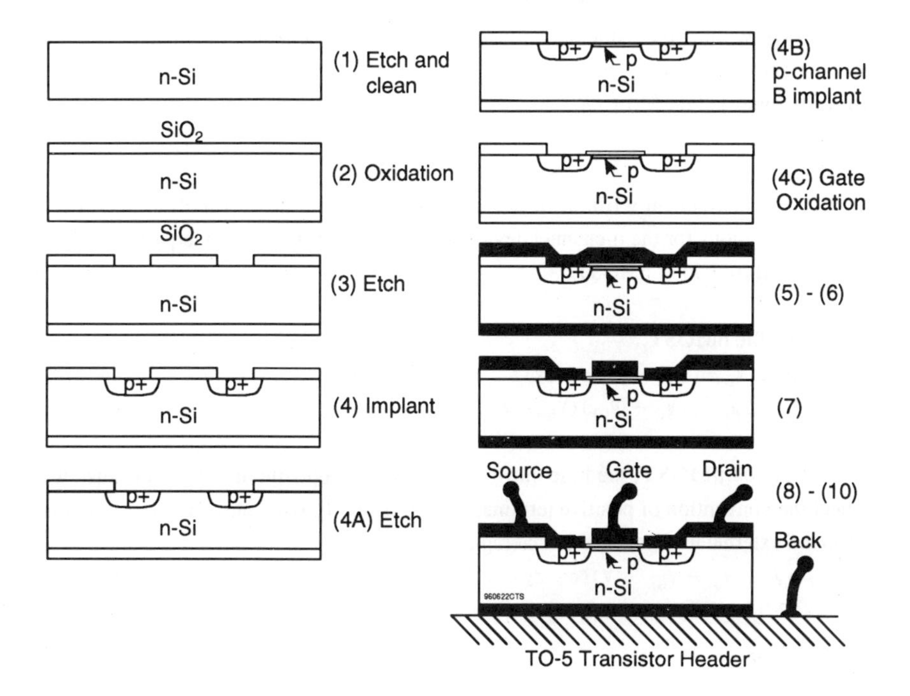

There are several possible ways to make the transistor by combinations of in-diffusion, ion implantation, epitaxial growth, outdiffusion, etc.. Any one combination is satisfactory. More than one additional step may be necessary for some combinations. The following is one sequence of steps.

(1)-(4) Same as Fig. 630.1 (1)-(4) except p-Si wafer changed to n-Si wafer, n+source/n+drain changed to p+source/p+drain, and phosphorus changed to boron for implantation of the p+ source and drain regions.

(4A) Etch oxide window for p-channel on top surface.

(4B) Ion implant a low dose of boron into n-Si through top oxide window (This gives the best dosage control but ion damage may give residual interface and oxide traps after activation-anneal.)

(4B1) Or diffuse a low concentration of boron into n-Si through top oxide window. (Very hard to control.)

(4C) Grow thin gate oxide in window, simultaneously activate implanted boron into active acceptor.

(5)-(10) Same as Fig. 630.1.

P641.1 By inspection without doing any algebra, write down the d. c. MOSFET equations for the p-channel device which are the pMOST counterpart to (643.1) and (643.2) of nMOST.

For the nMOST,

$$I_D = \mu_n C_O Z(V_G - V_{GT} - V)(dV/dy) \qquad (643.1N)$$

$$I_D = +(Z/L)\mu_n C_O[(V_G - V_{GT})(V_D - V_S) - (V_D^2 - V_S^2)/2] \qquad (643.2N)$$

For the pMOST, the hole current flows in the opposite direction, because we select the convention of positive terminal current is that flowing into a transistor region from the external lead attached to that region,

$$I_D = -\mu_p C_O Z(V_G - V_{GT} - V)(dV/dy) \qquad (643.1P)$$

$$I_D = -(Z/L)\mu_p C_O[(V_G - V_{GT})(V_D - V_S) - (V_D^2 - V_S^2)/2] \qquad (643.2P)$$

P641.2 Give the detailed derivation steps of the pMOST d.c. equations using the procedure in the text that gave nMOST d. c. equations listed in (643.1) and (643.2).

Longitudinal field:

$$-I_D = I_Y = \int_0^\infty J_{PY}(x)dx = \int_0^\infty q\mu_p P(x)E_Y dxZ$$

$$= \int_0^\infty q\mu_p P(x)(-dV/dy)dxZ = \mu_p(-dV/dy)Z\int_0^{X_C} qP(x)dx = \mu_p(-dV/dy)ZQ_C$$

where $Q_C = \int_0^{X_C} qP(x)dx$.

Transverse field:

Semiconductor charge:

$$Q_S = q\int_0^\infty \rho(x)dx = \int_0^\infty q[P - N + N_{DD}]dx$$

In the bulk: $\rho = q(P_B - N_B + N_{DD}) = 0$, so $N_{DD} = N_B - P_B$.

$$Q_S = \int_0^\infty q(P-N+N_B-P_B)dx = \int_0^\infty q[P(x)-P_B]dx - \int_0^\infty q[N(x)-N_B]dx = Q_P - Q_B$$

where $Q_P = \int_0^\infty q[P(x) - P_B]dx$ and $Q_B = \int_0^\infty q[N(x) - N_B]dx$.

By charge neutrality, $Q_G+Q_{OT}+Q_{IT}+Q_S = 0$ and using $Q_G = C_O(V_G-V-\phi_{MS})$,

$$C_O(V_G - V - \phi_{MS}) + Q_{OT} + Q_{IT} + Q_P - Q_B = 0.$$

$$Q_P = -C_O[V_G - V - \phi_{MS} + (Q_{OT} + Q_{IT} - Q_B)/C_O] = -C_O[V_G - V_{GT} - V]$$

where $V_{GT} = \phi_{MS} - (Q_{OT} + Q_{IT} - Q_B)/C_O$.

Using the assumption $Q_C \simeq Q_P = -C_O[V_G - V_{GT} - V]$, then,

$$I_D = -\mu_p(-dV/dy)Z\{-C_O[V_G-V_{GT}-V]\} = -\mu_pC_OZ(V_G-V_{GT}-V)(dV/dy). \quad (643.1P)$$

Then, perform the integration, we get

$$\int_0^L I_D dy = -\int_{V_S}^{V_D} \mu_pC_OZ(V_G - V_{GT} - V)dV \quad \text{which gives}$$

$$I_D = -(Z/L)\mu_pC_O[(V_G - V_{GT})(V_D - V_S) - {}^1\!/\!_2(V_D^2 - V_S^2)] \quad (643.2P)$$

P641.3 Use simple classical particle kinetics to show that the transverse electric field in the MOST is not a fundamental cause of mobility and drift velocity variation with the electric fields.

The reduction of mobility and drift velocity is due to irreversible energy loss by the channel conduction electrons (in nMOST) or holes (in pMOST) during their scattering and trajectory deflection by random lattice vibrations and randomly located impurity atoms and intrinsic defects. The variation (reduction) of the mobility and drift velocity (below its value predicted by mobility times electric field) with the applied electric field (reduction as the electric field increases) signifies that some of these energy loss pathways have higher loss rates (i.e. higher dissipation rate of the kinetic energy of the electrons gained from acceleration from the electric field) at higher electric fields. The power loss or energy dissipation rate which decreases the kinetic energy of the channel carriers (electrons in nMOST) must be balanced or equalled to the power gain or energy absorption rate by the channel electrons (or holes) from the applied electric field to accelerate the channel electrons (or holes) to higher kinetic energies. Classically, this power gain or power input is given by the J dot E product (current density vector dot electric field vector), **J•E**. Thus, for the MOST channel lying in the y-direction, the conduction current is flowing in the y-direction and the power gain is **J•E** $= J_Y \times E_Y$. Therefore, the perpendicular electric field, E_X, does not give input power and cannot be the fundamental cause.

The two-dimensional (and 3-D) picture is more complicated because the channel current is not confined entirely in the y-direction but has a small although significant x-component. Similarly, the two-dimensional (and 3-D) electric field distribution has also a small x-component. Thus, these two x-components will give a nonvanishing $J_X \times E_X$ to the input power gain or the rate of kinetic energy increase of the channel conduction electrons.

P642.1 Derive the effective trapped oxide charge density Q_{OT} given by (642.4).

This is the equivalent or effective areal density of the sheet of oxide charges located in an infinitesimally thin sheet at the SiO_2/Si interface which would have induced the same amount of energy band bending as the true physically distributed oxide volume charges located in the entire oxide film of thickness x_o. This one-dimensional formula can be written down without doing any algebra by taking advantage of the simple known one-dimensional answers of the parallel capacitor model. Thus, the charges induced in the Si surface space-charge layer by the volume distributed oxide charge with volume charge density of $\rho_{OT}(x)$ C/cm^3 is split between the Si and the gate conductor. The split follows the capacitor ratio of two capacitors between the gate and this layer of oxide trapped charges and the silicon and this layer of oxide trapped charges. The larger the capacitance between the two plates or the closer the two plates, the larger the induced charge on the plate by the sheet or plate of oxide charge. Thus, the negative charge induced in the Si by a layer of positive oxide charge located at x and x+dx is

$$dQ_{S\text{-induced}} = -[\rho_{OT}(x)dx]\times\{C(x_o-x)/[C(x_o-x)+C(x)]\}$$

which can be simplified by using the parallel capacitor formula,

$$C(x)=\varepsilon_o/x \text{ and } C(x_o-x)=\varepsilon_o/(x_o-x)$$

to give

$$dQ_{S\text{-induced}} = -\ [\rho_{OT}(x)dx]\times(x/x_o)$$

which can then be integrated through the thickness of the oxide from $x=-x_O$ to $x=0$ to give (642.4), $Q_{S\text{-induced}} \equiv Q_{OT}$. The students need to figure out why there is a sign difference, a typo, a difference in the direction of the x-axis or, or a different MOS orientation than the standard left to right used throughout the textbook.

P643.1 The d.c. drain current, I_D, of nMOST saturates to a constant value when the d.c. drain voltage, V_D, exceeds the d.c. gate voltage in excess of the gate threshold voltage, V_G - V_{GT}. The underlying physics is that the channel electron density is depleted to zero at the boundary of the reverse biased drain/channel junction when $V_D \geq V_G$ - V_{GT}. Prove the following statements mathematically for the nMOST with $V_X=V_S=0$ as the reference potential, using (642.2) for $Q_{C'}$ (the electron charge density or q times the electron number density)

$$Q_{C'} = C_o(V_G - V_{GT} - V) \qquad (642.2)$$

Use (643.1) for I_D and $dV(y)/dy = -E_y$

$$I_D = \mu_n C_o Z(V_G - V_{GT} - V)(dV/dy) \qquad (643.1)$$

Use (650.3) for $dV(y)/dy = -E_y$, and we get

$$dV(y)/dy = I_D/(\mu_n Z Q_N)$$

$$= \{[V_G-V_{GT})]^2-[V_G-V_{GT}-V_D]^2\}/\{2L[V_G-V_{GT}-V(y)] \qquad (650.3)$$

Use (643.4) for g_d, then

$$g_d = (Z/L)\mu_n C_o[V_G - V_{GT} - V_D] \qquad (643.4)$$

(a) The channel electron density is zero at the drain/channel boundary, y=L, where $V(y=L) = V_D$ when $V_D = V_G - V_{GT}$.

$Q_{C'}(V) = C_o(V_G - V_{GT} - V) \Rightarrow Q_{C'}(V=V_D=V_G-V_{GT})=0.$ Q.E.D.

(b) The channel electron density is zero at the drain/channel boundary, y=L, when V_D exceeds $V_G - V_{GT}$.

When $V_D>V_G-V_{GT}$, $Q_{C'}(V=V_D>V_G-V_{GT})=C_o(V_G-V_{GT}-V_D) < 0$

but $Q_{C'} \propto$ electron number = positive or zero. $\therefore$ $Q_{C'}=0$. Q.E.D.

(c) The channel electron density is the highest at the source/channel boundary, y=L, where V(y=0)=0.

$$Q_{C'}(V) = C_o(V_G-V_{GT}-V)$$
$$\Rightarrow Q_{C'}(V=0) = C_o(V_G-V_{GT}) > C_o(V_G - V_{GT} - V).$$ Q.E.D.

(d) The channel electron density at the source/channel boundary is not controlled by the drain voltage at all drain voltages.

$$Q_{C'}(V) = C_o(V_G - V_{GT} - V) \Rightarrow Q_{C'}(V=0)=C_o(V_G-V_{GT}) \neq f(V_D).$$ Q.E.D.

(e) The channel electric field or $E_y(y)$ = -dV(y)/dy is the lowest at the source/channel boundary, y=0 where V(y=0)=0.

$$I_D = \mu_n C_o Z(V_G - V_{GT} - V)(dV/dy) \qquad (643.1)$$
$$\therefore E_y = -dV/dy = -I_D/[\mu_n C_o Z(V_G - V_{GT} - V)]$$
$$\Rightarrow |E_y(y=0)| = I_D/[\mu_n C_o Z(V_G - V_{GT})]$$
$$< I_D/[\mu_n C_o Z(V_G - V_{GT} - V)].$$ Q.E.D.

(f) The channel electric field at the source/channel boundary, y=0, does depend on the drain voltage, but not linearly. What is the V_D dependence?

Given

$$E_y = -dV/dy = - I_D/[\mu_n C_o Z(V_G-V_{GT}-V)]$$
$$\therefore E_y(y=0) = - I_D/[\mu_n C_o Z(V_G-V_{GT})] = f(V_D)$$ Q.E.D.

V_D Dependence:

$$dV(y)/dy = I_D/(\mu_n Z Q_N)$$
$$= [(V_G-V_{GT})^2-(V_G-V_{GT}-V_D)^2]/\{2L[V_G-V_{GT}-V(y)]\} \qquad (650.3)$$
$$E_y(y=0) = -dV(y=0)/dy$$
$$= -[(V_G-V_{GT})^2-(V_G-V_{GT}-V_D)^2]/[2L(V_G-V_{GT})]$$

P643.2 The MOST d.c. current-voltage differential equation (643.1) and the d.c. terminal current-voltage equation (643.3) were derived from the drift current equation in the conduction channel between the source and the drain:

Current Density = (conductivity) times (electric field)

$J_y = \sigma_n E_y \equiv (q\mu_n N)\times(-dV/dy)$

Current = Current Density integrated over cross-sectional area

$-I_D = I_y = \int J_y dxZ$

and the solution of the above first order differential equation.

Show that this can also be obtained via ohm's law by adding the series resistances of each differential length (Δy) of the channel using:

Resistance of a differential length Δy = (resistivity) times Δy divided Area.

$\Delta R = (1/\sigma_n)\times\Delta y/\Delta xZ.$

Let R be defined = sum of all the ΔR of each thin Δx slice in parallel. Let us use the conductance which is the reciprocal of the resistance in order to be able to add the parallel conductances easily: $R = 1/G$. Then,

$G = \Sigma\Delta G = \Sigma(1/\Delta R) = \Sigma(\sigma_n \Delta xZ/\Delta y = (Z\mu_n/\Delta y)\times(\Sigma N\Delta x) \equiv (Z\mu_n/\Delta y)\times Q_{C'}$.

From ohm's law: $\Delta V = V(y) - V(y-\Delta y) = -I_y\times R = I_D\times(1/G)$.

$\therefore I_D = -I_y = \Delta V/R = G\times\Delta V = (Z\mu_n/\Delta y)\times(Q_{C'})\Delta V$

$= (Z\mu_n)\times(Q_{C'})\times(\Delta V/\Delta y)$

$= Z\mu_n Q_{C'}(dV/dy) \equiv$ (641.1C) on. p.544 FSSE. Q.E.D.

P644.1 The cut-off frequency of the MOST can be estimated from the low-frequency transconductance and the gate oxide capacitance without considering their dependence on the d.c. drain and gate voltage dependence (i.e. using $\approx g_m/2\pi C_o$). Its intrinsic delay can be estimated from the carrier transit time from the source to the drain, $\approx$ Length/Velocity = $L/\mu_n(V_D/L) = L^2/\mu_n V_D \propto 1/2\pi f_{cut\text{-}off}$. What is the cut-off frequency and the intrinsic delay of the nMOST's of 5-micron and 1-micron technologies given in section 644? How do these compare with the gain-bandwidth product obtained from the small-signal equivalent circuit model given in section 653, and the intrinsic delay computed from the charge-control model given in section 661?

5μ-technology: $Z/L=20/5=4$ and $x_o=1000A$. $C_oZL=33.6fF$. $g_{ms}= 400\mu S$.

1μ-technology: $Z/L=1/1 =1$ and $x_o=100A$. $C_oZL=3.36fF$. $g_{ms}=1000\mu S$.

$f_{cut\text{-}off} \approx g_m/(2\pi C_o ZL)$

$= 400\mu S/(2\times3.1416\times33.6fF) = 1.8947\times10^9 Hz = 1.9GHz$ 5μ-Technology.

$= 1000\mu S/(2\times3.1416\times3.36fF) = 25\times1.89\times10^9Hz = 47.3GHz$ 1μ-Technology.

$t_{intrinsic} = L^2/\mu_n V_D$

$= (5\times10^{-4})^2/(600\times5) = 8.333\times10^{-11}s = 83.33ps$ 5μ-Technology.

$= (1\times10^{-4})^2/(600\times5) = 8.333\times10^{-11}s/25 = 3.33ps$ 1μ-Technology.

Gain-bandwidth product $= (3/4\pi)(\mu_n/L^2)(V_G-V_{GT}) = (3/4\pi)t_{intrinsic}$ (652.9)

The difference of (3/4π) is due to the approximate estimate made in this problem. If $t_{trs}=(4/3)L^2/\mu^n(V_G-V_{GT})$ from (661.5) is used instead of $t_{intrinsic}$, the difference become 1/π.

P644.2 From the results in P644.1, show that for a load capacitance of 1pF, the switching speed of the nMOST is limited by charging the load capacitance and not by the intrinsic delay.

Given $C_{load} = 1pF=1000fF >> C_o=33.6fF$ (5μ-technology) or 3.36fF ∴

$t_{total\text{-}delay} \simeq t_{circuit} + t_{transistor}$

$\equiv t_{load} + t_{intrinsic}$

$= (g_m/2\pi C_{load})^{-1} + (g_m/2\pi C_o)^{-1}$

$= 2\pi \times (C_{load} + C_o)/g_m$

5μ-Technology

$= 2\pi \times (1000fF + 33.6fF)/400\mu S$

$= 2\pi \times (1000fF + \text{drop})/400\mu S$

$= 5\pi\times10^{-9}s = 15.7ns >> 83.3ps$

1μ-Technology

$= 2\pi \times (1000fF + 3.36fF)/1000\mu S$

$= 2\pi \times (1000fF + \text{drop})/1000\mu S$

$= 2\pi\times10^{-9}s = 6.3ns >> 3.3ps$ Q.E.D.

P644.3 A new and most unambiguous and easily measurable figure-of-merit of a MOST transistor and MOST technology was recently introduced by Mark Bohr and his engineers of the Intel Corporation (a former graduate student of Sah) at the 1994 IEEE International Electron Device Meeting (See the 1994 IEDM Conference Proceedings.) in a paper describing the Pentium-Pro 0.35micron, 4nm oxide technology. The definition was $t_{delay} = C_oAV_{DD}/I_{D\text{-}sat}$ where C_o is the gate oxide capacitance per unit area, $C_oA=\varepsilon_oA/x_o$, V_{DD} is the power supply voltage or designed voltage for the technology, A is the area of the gate, A=W×L and $I_{D\text{-sat}}$ is the measured drain saturation current at the power supply voltage. Note the powerful feature of this definition: it depends only on three easily and accurately measurable parameters, the gate oxide thickness, the power supply voltage and the drain saturation current. **Show** that this actually overestimated the transistor's intrinsic speed by almost a factor of two.

From $I_{DS} = (Z/L)(\mu_n C_o/2)(V_G - V_{GT})^2 \equiv I_{D\text{-sat}}$ (643.6)

$V_{DD} = V_{DS} = V_G - V_{GT}$

$\therefore t_{delay} = C_o WLV_{DD}/I_{D\text{-sat}} = 2\times L^2/\mu_n V_{DD}$ Intel Definition

From $t_{trs} = (4/3)\times L^2/\mu_n(V_G-V_{GT})$ (661.5)

$t_{0.9} = 0.96984\times L^2/\mu_n(V_G-V_{GT})$ (661.6)

From $I_{D\text{-sat}} = C_0 Z\Theta_{sat}(V_G - V_{GT} - V_S) = C_0 Z\Theta_{sat}(V_G - V_{GT})$ (684.6A)

$g_{m\text{-sat}} = C_0 Z\Theta_{sat}$ (684.6B)

$t_{delay\text{-}\Theta sat} = C_0 A/g_{m\text{-sat}}$

$\equiv L/\theta_{sat} = 1\times L^2/\mu_n(V_G-V_{GT})$ (684.6C)

The above comparison, regardless of the exact formula used, indicated that this 0.35micron and other more advanced technology since 1994-IEDM have reached its theoretical limit from random scattering of the electrons or holes in the channel by optical phonons which limits the velocity of the electrons or holes to their maximum allowed value, about 10^7cm/s, which is the highest speed the transistor can be switched or the signal can be transmitted from the input gate to the output drain by the signal carrying electrons or holes in the channel. The data of the Bohr paper is replotted on the next page from the viewgraph used by Sah in an invited talk in May, 1995. The figure shows two points. (i) The L^2 dependence of long-channel old technologies cannot be from constant-mobility long-channel theory such as (661.5) and (661.6) because the mobility would be too low, but from interconnect delay on the chip whose RC time delay is $\propto L^2$. (ii) The L dependence of the short-channel technologies (such as the 1994's 862-P6 for Pentium-Pro on the figure) demonstrate the speed limit due to velocity saturation predicted by (684.6A) and (684.6B). The datapoints using the gate or channel length (G symbols) rather than effective channel length (dots) and the Intel definition given above are almost exactly twice of that given by the velocity saturation theory, (684.6C).

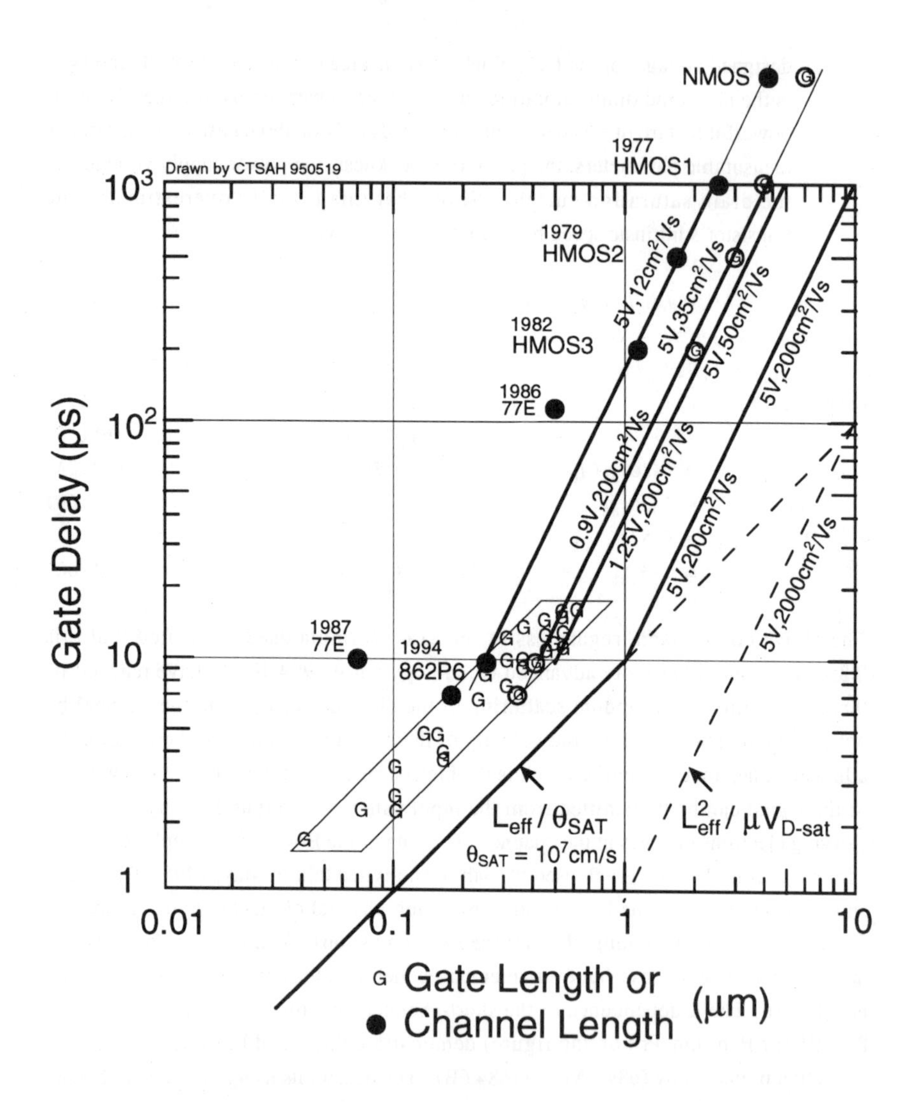

Fig.644.3 Channel length scaling limits. Experimental data and theoretical gate delay limits. From Viewgraph VG-09B presented by Chih-Tang Sah in "Physical Limits of CMOS Devices," an invited Lecture at the Sub-0.35μm CMOS Technology Workshop, Hsinchu, Taiwan, May 29, 1995.

P660.1 (corrected P654.2) The transconductance delay time (Fig.652.9), the transconductance and conductance cutoff characteristic frequencies (Figs.654.2 to 654.4), and the normalized channel transit time (Fig.661.2) are all given in normalized units along the ordinate (y-axis). They are not dimensionless variables frequently used by theorists. Discuss the basic device physics underlying the reason why they are not presented in dimensionless normalized unit but must use the normalization, L^2/μ_n, which has the dimension of Volt-second instead of second. The two-part answer is obvious if you examine the equations in the text that give the t's and ω's.

The reason is that the resulting equation are all dependent on voltages and hence can be easily scaled to lower voltages and shifted in voltage if the threshold voltage, V_{GT}, is not assumed zero. In addition, they can be normalized to the threshold voltage, V_{GT}, which would have exactly the same appearance as those given in the text. Some instructors, not appreciative of this choice of the figure's scale and ignorant of its underlying objectives, were led to believe of typographical errors and responded to students' questions as such. It is a cardinal example of using the Oriental 'one-glance' philosophy which must also be well-known among modern advertising agents to exploit the electronic advances applied to graphics art.

P660.2 (P663.1) In some parts of the text, C_L, is used. In other parts of the text, C_D is used. Give one or more practical examples illustrating this multiple usage.

C_L is used to compute the circuit delay of a logic stage to take into account the delay of the interstage interconnect line capacitance, which is denoted by C_L, the load capacitance and which may be simulated by a lump capacitance if the transistor is the final output transistor that drives the off-chip load. Thus, in these two circumstances, the C_L defines or characterizes two applications: (i) the inter-stage interconnect line oxide-capacitance for design optimization such as minimizing clock skew to attain lowest gate delay of a high transistor count integrated circuit chip and (ii) the maximum off-chip load capacitance that is allowed to the system designer to give the maximum speed or minimum guaranteed circuit operation speed and the nominal off-chip load capacitance to give the recommended or designed operating speed.

C_D is the drain capacitance or the equivalent input capacitance looking into the drain contact node with respect to a reference node. This does not include any load capacitance but does include all the parasitic capacitances arising from the extrinsic portion or geometric layout of a transistor, such as the overlap of the gate electrode over the oxide covering the n+drain surface, and the drain n+/p junction capacitance of a finite drain junction area. In the limit of ideal but practically unattainable layout (cannot have a n+drain junction with zero area with respect to the p-well or p-substrate although the thin-film SOI (Silicon on Insulator) technology almost approaches this ideal, there is no parasitic capacitance, in which case, C_D would be that of the intrinsic transistor which represents the actual charge stored in the n-channel as controlled or changeable by the applied drain voltage, $\partial Q_{CH}/\partial V_D$, with all other terminal voltages kept constant (known in circuit theory as short-circuit) or current constant (known in circuit theory as open-circuit). This is the 'intrinsic' capacitance of the intrinsic transistor for the one-lump network representation or equivalent circuit of the transistor which is voltage dependent and used in the analysis of the switching transient response of the transistor under a voltage pulse applied to the gate.

P660.4 (corrected P662.1) If the drain and gate supply voltages are dropped to 0.6V and the oxide reduced to 40A in a possible future technology as indicated by a recent (1989-1990) IBM paper and similar papers in the last few years, compute the device parameters required in Table 662.1. Discuss potential memory bit errors due to noise (i.e. fluctuations in voltage compared to the supply voltage.)

Assume $Z=W=0.15\mu m$, $L=0.05\mu m$, $x_o=40A$, $\mu_n=100cm^2/V\text{-}s$,
$V_G=V_D=0.6V$, $V_{GT}=0V$ and $T=300K$.

$C_o = \varepsilon_{ox}/x_o) = 3.9x8.854\times10^{-14}(F/cm)/40\times10^{-8}cm = 8.63\times10^{-7}F/cm^2$

$C_oLW = (8.63\times10^{-7}F/cm^2)\times(0.15\times0.05)\times10^{-8}cm^2=6.5\times10^{-17}F = 0.065fF$

$I_{DS\text{-}sat} = (Z/2L)\mu_nC_o(V_G-V_{GT})^2$

$= (0.15/2\times0.05)(100cm^2/V\text{-}s)(8.63x10^{-7}F/cm^2)(0.6-0)^2 = 46.6\mu A$

$g_{ms} = (Z/L)\mu_nC_o(V_G-V_{GT})$

$= (0.15/0.05)(100cm^2/V\text{-}s)(8.63x10^{-7}F/cm^2)(0.6-0V)$

$= 1.553\times10^{-4}S = 155.3\mu S$

Number = $(C_oV_G/q)(Area)$

$= [(8.63\times10^{-7}F/cm^2)(0.6V)/(1.6x10^{-19}C)](0.15x10^{-4}cm)(0.05\times10^{-4}cm)$

= 243 electrons on gate! (too small for memory)

$t_{trs} = (4/3)[L^2/(\mu_n(V_G-V_{GT})) = (4/3)(0.05\times10^{-4}cm)^2/(100cm^2/V\text{-}s\times0.6V)$
$= 5.6\times10^{-13}s = 0.56ps$

$t_{delay\text{-}\Theta sat} = L/\Theta_{sat} = 0.05\times10^{-4}/10^7 = 0.5ps \approx t_{trs}$

$E_{sw\text{-}sat} = (2/3)C_oLZ(V_G-V_{GT})^2$
$= (2/3)(8.63\times10^{-7}F/cm^2)(0.05\times10^{-4}cm)(0.15\times10^{-4}cm)(0.6-0V)^2$
$= 1.55\times10^{-17}J$

$kT = k\bullet300 = 4\times10^{-6}fJ = 4\times10^{-21}J$

$E_{sw}/kT = 1.55\times10^{-17}/4\times10^{-21} = 3.9\times10^3$

Since $E_{sw}/kT = 3900$, the fluctuations in voltage due to temperature (kT) should not affect the signal voltage and applications such as memory. Additional considerations or results of this example are the far reaching consequence of operating the MOST at as low a voltage as possible, such as that demanded in portable applications (including portable PC's heat dissipation and battery life). It shows that decreasing the operating voltage would decrease quadratically the power dissipation and proportionally increase the battery life in portable applications, however, the speed is reduced also at lower voltages. The importance of the threshold voltage becomes increasingly important because the speed is proportional to the applied voltage offset by the threshold voltage.

P660.13 (corrected P652.13) Why are the transconductance, g_m, given in (643.7) and (652.1) different?

$$g_m = (\partial I_D/\partial V_G)|_{V_D} = (Z/L)\mu_n C_o V_D \qquad (643.7)$$

$$g_m = g_{m0}/(1 + j\omega t_{gm}) \qquad (652.1)$$

$$g_{m0} = (\partial I_D/\partial V_G)|_{V_D} = (Z/L)\mu_n C_o V_D \qquad (652.2)$$

It is obvious that when we started to study the d.c. characteristics in section 643, we had not attained the concept of small-signal sinusoidal response which was studied in a later section, 652. Thus, for notation simplicity, we used g_m instead of g_{m0} in the definition and expression of the differential transconductance, given by (643.7). And when we analyzed the small-signal sinusoidal response in the later section, 652, we realized that we need to use a new notation for the low-frequency small-signal transconductance, g_{m0}, so that we can still use the symbol, g_m, for the sinusoidal steady-state small-signal transconductance. Strictly, we would have used G_m but the IEEE notation convention has used the capitalized symbol for external circuit parameters.

But, the most important two distinctions of the differential g_m in (643.7) and the sinusoidal small-signal g_m in (652.1) are the different measurement conditions or definitions, and the difference in basic physics involved. Both of these are using the source as the common or reference and the substrate and well terminals are all grounded to the source also. The differential g_m in (643.7) is measured by reading the change of d.c. drain current, ΔI_D, from a change of the d.c. voltage applied to gate, ΔV_G, i.e. $\Delta I_D = I_D(V_G+\Delta V_G) - I_D(V_G)$. There is no time involved or measurement speed involved since it is d.c. and the measurement time is as long as needed to be able to measure the change of drain current to the desired accuracy. However, the small-signal g_m in (652.1) is obtained by measuring the small-signal sinusoidal drain current (both amplitude, rms I_d or peak $I_{d\text{-}peak}$ and the phase, θ) due to an applied small sinusoidal gate voltage superimposed onto the d.c. gate voltage, $V_G + V_{g\text{-}peak}\exp(j\omega t)$ or $V_G + V_{g\text{-}peak}\sin(\omega t)$ or in the traditional electrical engineering's RMS notation, V_g for the rms value, instead of the peak value, $V_{g\text{-}peak}$, then the voltage is $V_G + V_g$. Then, the small-signal transconductance that is measured is $g_m = I_d/V_g$ which will decrease with increasing measurement frequency following the $1/(1+j\omega t_{gm})$ dependence given in (652.1) where t_{gm} is given by (652.4).

The fundamental physics difference is that the differential transconductance is a d.c. parameter which does not involve a measurement time as long as it is long enough to give the desired accuracy. Thus, signal delays due to the finite velocity of the signal carrying particles (electrons in nMOST and holes in pMOST) and due to storing and removing these current-carrying charges at some storage locations are irrelevant because the measurement time can be made as long as needed to allow I_D to change from one steady-state D.C. value at V_G to another at $V_G+\Delta V_G$.

However, the small-signal measurement of the transconductance, which is also a change of the drain current per unit change of the gate voltage, is measured in a finite time, if it is a transient measurement involving an abrupt gate voltage step with the drain current rising towards its final value with a delay rather than instantaneously because the channel carriers have to travel from the source to the drain end of the channel to produce the change of the drain current. Thus, if the drain current is measured at a very short time soon after the gate voltage step, then the drain current would not have reached its final value, and this is accounted for by the exponential time constant by replacing $j\omega t_{gm}$ by st_{gm} where s is the Laplace transform variable or by just solving the small-signal time-dependent differential equation directly.

Similarly, if the small-signal transconductance is measured sinusoidally which usually is measured over many periods of the sinusoidal voltage, known as sinusoidal steady-state, then there is a difference in phase when small-signal sinusoidal waveforms of $v_g(t)$ and $i_d(t)$ are displayed as a function time on an oscilloscope. This phase difference represents the delay of the signal-carrying carriers which must travel from the source to the drain in the channel to change the drain current.

Thus, the basic physics difference of the differential and small-signal transconductances is that the small-signal transconductance contains one additional physics, namely the transit time delay of the electrons or holes passing through the channel length L from the source to the drain which is not measured in the differential transconductance which only measures the electron or hole mobility and the oxide thickness that determines the transverse electric field which determines the concentration or number of electrons or holes in the channel.

P660.14 (corrected P652.14) Can the transconductance in (643.7), (652.1) and (652.2) be used at any d.c. drain voltages, V_D, regardless of what the d.c. gate voltage is? Why?

No. They apply only when V_D is smaller or equal to V_G-V_{GT} or in the non-saturation region of the I_D-V_D characteristics. The underlying physics is that when V_D is larger than V_G-V_{GT}, then the excess, $\Delta V_D \equiv V_D - (V_G-V_{GT})$ will all appear through the thickness of the space-charge layer of the drain/channel junction and will not change the potential barrier at the source/channel junction. And, only the change of the source/channel potential barrier will change the channel current via changing the amount of carriers supplied by the source. Thus, the transconductance will reach a constant value given by $V_D = V_G - V_{GT}$ and no higher if V_D is increased further while V_G is kept constant. (This is the ideal 1-D long channel solution. In shorter channels, the effective length of the channel would be shortened by the penetrating drain/channel space-charge layer into the channel, resulting in a slowly increasing drain current or transconductance with additional drain voltage increase over V_G-V_{GT}.)

P670.1 Refer to Fig.670.2, show that the MOSFET symbol is incorrect and misleading when there are three arrows, one each for the drain, source and substrate contract. Realize the physical structure that can be represented by such a three arrow symbol.

In Fig.670.2, there are two sets of three-arrow (or three-diode) MOST circuit symbols one for each of the four MOSTs, p-channel and n-channel, and enhancement and depletion. The FCI set (x,y)=(6,1→4) uses thin and thick line to distinguish respectively enhancement (no channel conduction at V_G=0 and channel must be induced by an applied V_G) and depletion (channel conduction at V_G=0 which can be cutoff by applying a V_G to deplete the carriers in the channel). The RCA set (x,y)=(6,1'→4') uses three broken lines and one solid but thin line to distinguish respectively enhancement and depletion. Both of these are confusing. The thin line for the enhancement pMOST and nMOST in the FCI set implies conduction lower (or thinner conductance) than the thick line for the two depletion MOSTs while the intent of the thinner line is to show no conduction. The three broken lines for the enhancement nMOST and pMOST in the RCA set implies three partially extended channels not-connected to each other while the intent is for no channel at all. Nevertheless, the three-diode symbol is strictly correct for both d.c. bias and a.c. signal analyses.

The correct while unambiguous set would be the hybrid of the three-arrow or three-rectifying-diode FCI and RCA sets with the RCA set modified as follows. The depletion pMOST and nMOST are represented by FCI's (6,3) and (6,4) whose thickline represents the existing channel that must be depleted by a proper polarity applied V_G. The enhancement pMOST and nMOST are represented by RCA's (6,1') and (6,2') whose three broken lines are reduced to zero length to show that there is no partial-length channel. This then allows the symbol to be extended to composite MOST's with common gate but disconnected channels or special MOST structures with partial-length channel such as those used in cascode amplifiers in receivers. This best-choice set of four symbols are shown in Fig.P670.1 given below. Physical structure realizations are given in Fig.672.21 and additional ones can be constructed by the students.

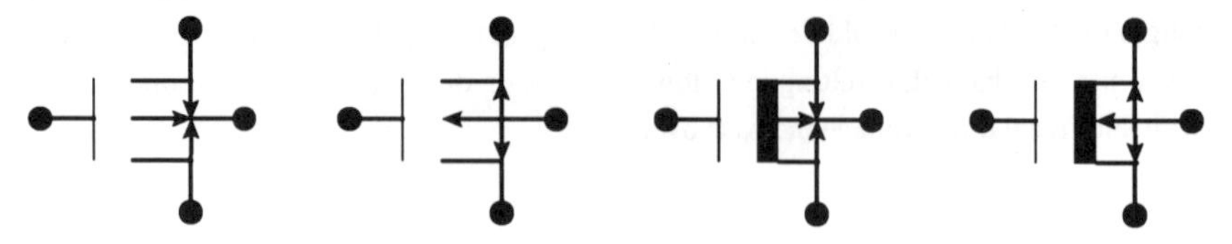

Chapter 7
BIPOLAR JUNCTION TRANSISTOR AND OTHER BIPOLAR DEVICES

OBJECTIVES

Review of invention history
Definition of device structure parts and terms
Typical fabrication sequence of a BJT
D. C. characteristics

- Two-diode representation
- Experimental data (from production data sheets)
- Mathematical derivation, Shockley BJT and SNS BJT equations
- Original and extended Ebers-Moll equations
- Two-port nonlinear d. c. network representation
- Lumped d. c. model of real multi-dimensional BJT
- Material and structure dependences of d. c. parameters
- Bias dependences of d. c. parameters
- Collector multiplication and negative resistance

Small-signal characteristics

- The small-signal conditions
- Exact analysis for small-signal equivalent circuit models
- Common-Base small-signal Tee (CBss-Tee) models
- Maximum frequency of oscillation of BJT (the Gibbons frequency)
- Common-Emitter small-signal Hybrid-Pi (CEss-Hπ) models
- Common-Emitter current gain, cutoff frequency, bandwidth

Large-signal switching characteristics

- Diffusion and charge-control equations
- Common-Base Large-signal BJT switching transient
- Common-Emitter Large-signal BJT switching transients
- Comparison of CB and CE BJT switching transients
- Speeding up BJT via technology

Circuit applications of BJT

Heterostructure bipolar junction transistors (HBJTs)

The four-layer PNPN devices
PNPN triode (SCR) characteristics
MOS-SCR
Latch-up in CMOS revisited

P710.1 Determine from the literature the basic reasoning underlying the first analysis of minority carrier injection and diffusion made by Shockley in 1946. How was this different from earlier work by British and German research scientists and book authors?

The answers to this question require a reading of Shockley's mid-70 historical review article on the invention of transistors in a special IEEE Proceeding issue where he showed the figures in his 1946 patent notebook entry at the Bell Telephone Laboratories on his attempt to analyze the minority diffusion. The motivation was obviously to find out how to characterize and delineate the origin of the current flowing in a semiconductor diode or two-terminal structure which consists of a p-type region and an n-type region. The basic underlying reasoning was that the n-type region has much higher electron concentration than the p-type region, therefore, there is a concentration gradient, dn/dx, associated with which must be a diffusive flow and electrical current. Then, Shockley sketched the energy band diagram of the semiconductor versus distance perpendicular to the planes of the p-type and n-type regions or layers and noticed that the electrons flowing diffusively from the n-type region to the p-type region must move against or overcome a potential hill or potential barrier, so the word injection was used.

This was in contrast to the earlier work by British and German researchers whose work were summarized in several semiconductor and metal physics books and original articles cited in these books [199.2, 199.3, especially 599.1, 599.2, 599.3] which were focused solely on thermionic emission (by heating one part of the two material sandwich or diode) or field emission (by applying an electric field or a voltage between the two leads of the diode) emission over a similar repulsive potential barrier between two dissimilar materials, such as metal/vacuum and metal/semiconductor barriers. These were not minority carriers after they are injected over the barrier to the other side. In the case of thermionic emission of electrons from the metal into the vacuum, there is no material in the vacuum to classify the electrons as majority or minority since there is no hole-quasi-particles in vacuum. In the case of field or applied voltage caused emission of metal electrons into the semiconductor in the metal/semiconductor rectifiers, the semiconductor was n-type since the concept of holes was in its infancy and not so common place as today, so the injected electrons in the n-type semiconductor are still majority carriers. And the majority carrier current is dominated by drift while the majority-carrier diffusion current is negligible since the majority carrier concentration is essentially spatially constant.

The students are strongly urged to read the historical articles and books just cited to gain added appreciation on the eventually developed theories we use today in sophomore-junior textbooks, to know the roots of the current manufacturing practice, to write their own summary on transistor evolution and to tell their own story of the history of computer chips to the laymen.

P731.1 By inspection without doing any algebra, write down the d. c. equations of the n/p/n BJT transistor using the two-diode model.

For the p/n/p BJT transistor, by inspection of the general d.c. equivalent circuit in Figure 731.1 (c), we can write down the following pair of equations using Kirchoff's current law without doing any algebraic derivation.

$$I_C = I_{CD} - \alpha_F I_{ED} \qquad (731.1)$$
$$I_E = I_{ED} - \alpha_R I_{CD} \qquad (731.2)$$

For the n/p/n BJT transistor, by inspection of the general d.c. equivalent circuit in Figure 731.2 (c) we can write down the following pair of equations using Kirchoff's current law without doing any algebraic derivation.

$$I_C = -I_{CD} + \alpha_F I_{ED}$$
$$I_E = -I_{ED} + \alpha_R I_{CD}$$

This problem demonstrates the power of basic circuit theory in applications to highly nonlinear elements, giving simple but very general equations that are particularly useful in engineering design of million-transistor integrated circuits via the use of compact and yet accurate nonlinear device (diodes and transistors) models to reduce the computer simulation time.

P733.1 Obtain the qualitative criteria on the fundamental material parameters (energy gap, mobility, lifetimes, and dopant impurity concentration) for the Shockley diode current to dominate over the SNS current, under reverse and forward biases, and at low and high temperatures (in a BJT).

There are two approaches that may be taken to compare the Shockley diode current with the SNS current. (1) Compare J_E(Shockley) with J_E(SNS) where J_E(Shockley) consists of three terms, $J_E(\text{Shockley}) = J_N(O_E,V_{EB}) + J_P(O_B,V_{EB}) + J_P(0_B,V_{CB})$ and similarly with the collector current. (2) Compare the Shockley and SNS components in the compact form of the emitter and collector currents. In both cases, a comparison may be made of the terms directly or by calculating the voltage at which the two terms become equal similar to the comparison of the Shockley and SNS components of the p/n junction diode current in Chapter 5 (Section 535) which defined an injection threshold voltage and a simple formula.

$$J_E = +(j_E+j_{EB}+j_B)[\exp(qV_{EB}/kT)-1] - (\alpha_B j_B)[\exp(qV_{CB}/kT)-1] \qquad (733.19)$$
$$J_C = -(\alpha_B j_B)[\exp(qV_{EB}/kT)-1] + (j_C+j_{CB}+j_B)[\exp(qV_{CB}/kT)-1] \qquad (733.20)$$

(1) At forward configuration, $V_{EB} > 0$ and $V_{CB} < 0$.

The collector current (733.20) reduces to:

$$J_C = -(\alpha_B j_B)[\exp(qV_{EB}/kT)-1]$$

Therefore, the SNS component is negligible in the collector current in the forward configuration since the collector/base junction is reverse-biased.

The emitter current (733.19) reduces to:

$$J_E = +(j_E+j_{EB}+j_B)[\exp(qV_{EB}/kT)-1] \text{ where}$$
$$j_E = (qD_EN_E/L_E)\text{ctnh}(X_E/L_E) \qquad (733.21A)$$
$$j_{EB} = (qn_iX_{EB}/\tau_{EB})[\exp(qV_{EB}/2kT)-1]/[\exp(qV_{EB}/kT)-1] \qquad (733.21B)$$
$$j_B = (qD_BP_B/L_B)\text{ctnh}(X_B/L_B) \qquad (733.21C)$$

Separating out J_E into the Shockley and SNS components:

$$J_E = J_E(\text{Shockley}) + J_E(\text{SNS})$$
$$J_E = (j_E+j_B)[\exp(qV_{EB}/kT)-1] + j_{EB}'[\exp(qV_{EB}/2kT)-1]$$

where

$$j_{EB}' = (qn_iX_{EB}/\tau_{EB})$$

One method to compare the relative magnitudes of the Shockley and SNS components is to obtain the injection threshold voltage by equating the Shockley and SNS components (identical to the method used to obtain Eq. (535.7)) and to determine how this threshold voltage varies as the material parameters are varied.

$$(j_E+j_B)[\exp(qV_{EB}/kT)-1] = j_{EB0}[\exp(qV_{EB}/2kT)-1]$$

$$(j_E+j_B)[\exp(qV_{EB}/kT)\quad] \simeq j_{EB0}\exp(qV_{EB}/2kT)$$

$$V_{EBIT}\ \text{(EB Injection Threshold)} = (2kT/q)\log_e[j_{EB0}/(j_E+j_B)]$$

$$= (2kT/q)\times\log_e\{(qn_iX_{EB}/\tau_{EB})/[(qD_EN_E/L_E)\text{ctnh}(X_E/L_E)+(qD_BP_B/L_B)\text{ctnh}(X_B/L_B)]\}$$

This can be simplified by $\text{ctnh}(z) \simeq 1/z$ since z is small. Also, use the Einstein relation, $D/\mu = kT/q$, then

$$V_{EBIT} = (2kT/q)\log_e\{(qn_iX_{EB}/\tau_{EB})/[(kT\mu_EN_E/X_E)+(kT\mu_BP_B/X_B)]\}$$

Use the mass action law, $N_E = n_i^2/P_E \simeq n_i^2/N_{AAE}$ where N_{AAE} is the dopant impurity concentration of the p+ emitter and $P_B = n_i^2/N_B \simeq n_i^2/N_{DDB}$ where N_{DDB} is the dopant impurity concentration of the n-type base. Then,

$$V_{EBIT} = (2kT/q)\log_e[(qn_iX_{EB}/\tau_{EB})/[(kT\mu_En_i^2/N_{AAE}X_E)+(kT\mu_Bn_i^2/N_{DDB}X_B)]]$$

Finally, use the equation for the space charge layer thickness of the emitter/base junction which is identical to (523.19).

$$X_{EB} = \sqrt{[2\varepsilon_S(V_{bi}-V_{EB})/qN_{DDB}]}.$$

Then, we can substitute this into the above and get

$$V_{EBIT} = (2kT/q)\times$$
$$\log_e\{[q\sqrt{[2\varepsilon_S(V_{bi}-V_{EB})/qN_{DDB}]}/\tau_{EB}]/[(kT\mu_En_i/N_{AAE}X_E)+(kT\mu_Bn_i/N_{DDB}X_B)]\}$$

Note that the energy gap affects n_i: $n_i = [N_CN_V\times\exp(-E_G/2kT)]^{1/2}$.

(1A) E_G: As E_G increases, n_i decreases, V_{EBIT} increases, and Shockley component decreases (less dominant). (This simple comparison does not take into account the temperature and doping concentration dependences of E_G.)

(1B) μ_E and μ_B: As the mobility increases, V_{EBIT} decreases, and Shockley component increases (more dominant). (This comparison does not consider the temperature and doping concentration interdependences.)

(1C) τ_{EB}: As τ_{EB} increases, V_{EBIT} decreases, and Shockley component increases (more dominant).

(1D) N_{AAE} and N_{DDB}: As N_{AAE} and N_{DDB} increase, V_{EBIT} increases, and Shockley component decreases (less dominant).

(1E) T: As T increases, V_{EBIT} increases, and Shockley component decreases (less dominant).

(2) At reverse configuration, $V_{EB} < 0$ and $V_{CB} > 0$.

The emitter current (733.19) reduces to:

$$J_E \approx -(\alpha_B j_B)[\exp(qV_{CB}/kT)-1]$$

Therefore, the SNS component is negligible in the emitter current in the reverse configuration since the emitter/base junction is reverse-biased.

The collector current (733.20) reduces to:

$$J_C = (j_C+j_{CB}+j_B)[\exp(qV_{CB}/kT)-1]$$

where

$$j_C = (qD_C N_C/L_C)\mathrm{ctnh}(X_C/L_C) \quad (733.21E)$$

$$j_{CB} = (qn_i X_{CB}/\tau_{CB})[\exp(qV_{CB}/2kT)-1]/[\exp(qV_{CB}/kT)-1] \quad (733.21F)$$

$$j_B = (qD_B P_B/L_B)\mathrm{ctnh}(X_B/L_B) \quad (733.21C)$$

The equations are identical to those considered above. Therefore, we can write the collector/base injection threshold voltage, V_{CBIT} as:

$$V_{CBIT} = (2kT/q)\times \log_e\{[q\tau_{CB}^{-1}\surd[2\varepsilon_S(V_{bi}-V_{CB})/qN_{DDB}]]/[(kT\mu_C n_i/N_{AAC}X_C)+(kT\mu_B n_i/N_{DDB}X_B)]\}$$

(2A) E_G: As E_G increases, n_i decreases, V_{CBIT} increases, and Shockley component decreases (less dominant). (This simple comparison does not take into account the temperature and doping concentration dependences of E_G.)

(2B) μ_C and μ_B: As the mobility increases, V_{CBIT} decreases, and Shockley component increases (more dominant). (This comparison does not consider the temperature and doping concentration interdependences.)

(2C) τ_{CB}: As τ_{CB} increases, V_{CBIT} decreases, and Shockley component increases (more dominant).

(2D) N_{AAC} and N_{DDB}: As N_{AAC} and N_{DDB} increase, V_{CBIT} increases, and Shockley component decreases (less dominant).

(2E) T: As T increases, V_{CBIT} increases, and Shockley component decreases (less dominant).

P733.2 Obtain the Shockley diode current coefficient, J_1, for the collector-base junction of a n+/p/n/n+ BJT which has a buried n+ collector and whose n-type collector is thin (given by X_C). Let the p-base/n-collector junction be located at x=0 and the boundary condition at the n/n+ interface be J_P=0 or dP/dx=0. Discuss the result at which the reverse bias is large enough such that the collector/base junction space-charge (depletion) layer becomes equal to the n-layer thickness ($x_{cb\text{-}n}=X_C$), which is known as punch-through.

We assume an constant-abrupt p-base/n-collector junction with spatially constant N_{AA} and N_{DD} dopant impurity concentration. We also assume that N_{DD+} of the n+buried collector is much greater than N_{DD}. Thus, the depletion solution of the space-charge layer thickness on two sides of the p-base/n-collector junction can be written using the previous solutions for the BJT, except one difference, J_P=0 at the n/n+ interface. However, this can be written down without solving the diffusion equation of holes in the thin n-collector of thickness X_c by noting that the diffusion current components given by the solutions in section 733, such as (733.6A) and (733.6B) are in the form of hyperbolic functions from taking the spatial derivative of the hole (minority carrier) distribution given by (733.5). The solution for P(x) in the X_c collector layer will have the form of (733.5) except that dP/dx = 0 at x=X_c so unity at x=0, so that the sinh must be replaced by cosh so,

$$P(x) - P_N = P_N[\exp(qV_{BC}/kT)-1]\cosh[(\Delta X_C-x)/L_p]/\cosh[(\Delta X_C/L_p)] \quad (733.5A)$$

$$J_P(x) = -qD_p dP/dx$$

$$J_P(x) = q(D_pP_N/L_p)[\exp(qV_{BC}/kT)-1]\sinh[(\Delta X_C-x)/L_p]/\cosh[(\Delta X_C/L_p)] \quad (P733.2A)$$

$J_P(x=\Delta X_C) = 0$ Check boundary condition at n/n+ interface.

$$J_P(0) = q(D_pP_N/L_p)[\exp(qV_{BC}/kT)-1]\tanh[(\Delta X_c/L_p)] \quad \text{(P733.2B)}$$

$$\therefore \quad J_1 = q(D_pP_N/L_p)\tanh[(\Delta X_c/L_p)] \quad \text{(P733.2C)}$$

Here, P_N is the equilibrium hole concentration in the thin n-collector given by $P_N=n_i^2/N_{DD}$ and x=0 is the edge location of the collector/base junction space-charge layer in the thin n-collector. (We do not use the coordinate suggested by the problem so the solution formulas given below are in simpler forms. The students may try the suggested reference location for x=0 to practice the derivation or verify the solutions such as (733.5A) by solving the differential equation (733.3) and by finding the two coefficients in (733.4). Using the depletion approximation for the reverse-biased base/collector junction $V_{CB}=-V_{BC}\geq 0$ we have

$$\Delta X_C = X_C - x_n \quad \text{(P733.2D)}$$

$$x_p = x_{pn}N_{DD}/(N_{AA}+N_{DD}) \quad \text{(523.23C) (P733.2E)}$$

$$x_n = x_{pn}N_{AA}/(N_{AA}+N_{DD}) \quad \text{(523.23C) (P733.2F)}$$

$$x_{pn} = [2\varepsilon_s(V_{CB-bi}+V_{CB})/qN_M]^{1/2} \quad \text{(531.1) (P733.2G)}$$

$$N_M = N_{AA}N_{DD}/(N_{AA}+N_{DD}) \quad \text{(P733.2H)}$$

When the revserse BC bias voltage is large enough that $x_n=X_C$ or $\Delta X_C=0$, the space-charge layer **punch-through** condition is reached. In this case of a LO-HI or n/n+ junction, holes cannot injected over the n/n+ potential barrier (because we assumed $J_P=0$ at the n/n+ interface, or equivalently, $N_{DD\text{-}n+} >> N_{DD}$. This is verified by the solution for J_1, (P733.2C), which shows that J_1 is zero which represents the fact that the quasi-neutral region of the thin n-collector is completely depleted of electrons and holes, hence there are no minority carrier or holes to be generated via the SRH model. But, holes are still generated in the depleted space charge layer which is given by J_2 in the SNS equation, therefore, the total CB junction leakage current is still finite, not zero, and in fact continually dominated by the SNS generation of electrons and holes at traps in the energy gap located in the depleted collector-base space-charge layer. For a properly designed n+/p/n/n+ BJT, we would have $N_{DD}<<N_{AA}$ in order to minimize the Early effect, namely, the thinning of the quasi-neutral base layer thickness by the reverse voltage applied to the BC junction discussed in section 738 and given by (738.2B). Then, with $N_{DD}<<N_{AA}$, so that $x_n>>x_p$ or the CB space-charge layer is mostly in the thin n-type collector layer. In which case, the SNS leakage current will no longer increase with the CB reverse bias when $x_p=X_C$ or $\Delta X_C=0$ or punch-through is reached because the generation volume in the space-charge layer has reached a constant and can no longer increase with the CB applied reverse voltage as indicated by (P733.2G) above.

P736.1 Derive the current equation of the underlap diode of a p+/n/p BJT whose base-collector junction depth is X_{BJ} as indicated in Fig.736.1. Assume that the base-metal contact to the n-base surface is ideal (i.e. zero contact resistance or infinite recombination-generation rate). What happens to the current when the collector/base junction bias is so high that its space-charge layer in the base layer side becomes equal to X_{BJ} and explain any limitations on the current?

This problem illustrates a detrimental punch-through effect, in contrast to that in problem P733.2 which showed a beneficial punch-through effect. This difference comes from the zero-current or zero-recombination/zero-generation of electrons and holes at the n/n+ interface in the previous problem and the infinite recombination-generation rate at the n-base/metal contact. The mathematics can again be avoided by writing down the solution by inspection of the existing solutions as in P733.2. The overlap diode is a p-collector/n-base/metal diode with a n-base thickness comparable to the minority carrier (holes) diffusion length in the n-base, $X_{BJ} \sim L_p$. Assuming again a constant-abrupt p-collector/n-base junction with constant N_{AA} in the collector and N_{DD} in the base, then the Shockley ideal diode current for this **thin-base** diode (The term, narrow-base, is seriously out-of-date and causes costly confusion.) is precisely that given by (733.6A) derived for the emitter current component of the p/n/p BJT of holes flowing from the p-emitter into the n-base towards a zero-biased n/p base/collector junction. The boundary condition of zero bias, and hence the carrier concentrations have their equilibrium value, is identical to the infinite recombination-generation rate condition assumed for the n-base/metal junction. Thus, this component of the Shockley ideal diode current is given by (733.6A) below with appropriate change of symbols

$$J_P(0_B) = (qD_pP_N/L_p)[\exp(qV_{EB}/kT)-1]\mathrm{ctnh}(X_B/L_p) \qquad (733.6A)$$

$$J_{P-ov}) = (qD_pP_N/L_p)[\exp(qV_{CB}/kT)-1]\mathrm{ctnh}(\Delta X_{BJ}/L_p) \qquad (P736.1A)$$

where the symbol definitions and their equations are very much like those in problem P733.2, with again only one change of symbol, i.e. replacing X_C by X_{BJ}. These are:

$$P_N = n_i^2/N_{DD} \qquad (P736.1B)$$

$$V_{CB-bi} = (kT/q)\log_e(N_{DD}N_{AA}/n_i^2) \quad \text{(P736.1C)}$$
$$\Delta X_{BJ} = X_{BJ} - x_n \quad \text{(P736.1D)}$$
$$x_p = x_{pn}N_{DD}/(N_{AA}+N_{DD}) \quad \text{(523.23C) (P736.1E)}$$
$$x_n = x_{pn}N_{AA}/(N_{AA}+N_{DD}) \quad \text{(523.23C) (P736.1F)}$$
$$x_{pn} = [2\varepsilon_s(V_{CB-bi}+V_{CB})/qN_M]^{1/2} \quad \text{(531.1) (P736.1G)}$$
$$N_M = N_{AA}N_{DD}/(N_{AA}+N_{DD}) \quad \text{(P736.1H)}$$

The divergence of the Shockley ideal current component given by (P736.1A) is evident when the punch-through condition is reached, $\Delta V_{BJ}=0$, from large reverse bias V_{CB} that increases x_{pn} and makes x_n equal to X_{BJ}. This is represented by $\text{ctnh}(\Delta X_{BJ}/L_p) = \text{ctnh}(0/L_p) = 1/\tanh(0/L_p) = 1/0 = \infty$ which makes $J_{P\text{-}ov} = \infty$. With increasingly smaller dimension and thinner layers to attain higher transistor density and higher speed, this divergence of the collector junction leakage current is real, not just academic exercise, and must be taken into account during the design.

The current of course will not become infinite because it is limited by the contact, interconnect, and lead resistances, and power supply current limit, but the current density will be so high and power dissipation so large (high reverse voltage) that irreversible damage by melting or by burning open the interconnect would occur.

P740.1 An electron device has a current voltage relationship of $i = Av^n$ down to $v=0$ where $0 < n < 1$. What is the condition for small-signal approximation to be valid?

Expand $i = Av^n$ in Taylor series at some voltage V using the notation $v=V+\delta v$ where $\delta v \ll V$. Then,

$$i = Av^n = A\{V^n + [\delta v/1!](dv^n/dv) + [(\delta v)^2/2!](d^2v^n/dv^2) + \ldots\}$$

Small-signal approximation is valid when the first nonlinear term, $(\delta v)^2$ is much smaller than the linear term or

$$|\delta v| \ll |2(dv^n/dv)/(d^2v_n/dv^2)|$$
$$= |2(nv^{n-1})/[n(n-1)v^{n-2}]|$$
$$= |2v/(n-1)|$$
$$= |2V/(n-1)|$$

Appendix
TRANSISTOR RELIABILITY

OBJECTIVES

- Delineation of failure origins: electronic and protonic traps.
- Mathematical analysis of MOST and BJT failures.

SUMMARY

This appendix contains background discussions, problems and solutions on transistor reliability including analysis for 10-year operation time-to-failure (TTF_{OP}). Four topics are discussed: interface traps, oxide traps, acceptor hydrogenation, and electromigration. Backgrounds of each topic are first presented which supplement the elementary descriptions given in FSSE (Fundamentals of Solid-State Electronics) and FSSE-SG (Study Guide). References, problems and solutions are numerically grouped to each subject by section number and placed at the end of the appendix. The primary and secondary literature references facilitate further investigations by the students and readers. Some of the problems given here were assigned in an introductory graduate level transistor reliability course and in an advanced graduate level transistor device and material physics course taught by the author and Professors Arnost Neugroschel and Toshi Nishida, during 1988 to 1996 at the University of Florida. Much of the background materials are derived from recent research investigations of this author with his graduate students and these two professorial colleagues.

The presentation format adheres to that of FSSE and the first 100 pages of this solution manual. The background sections are numbered by 9mn where m,n=0,1,.... are main and subsections numbers. The equations are numbered by (9mn.ee) and the figures are numbered by Fig.9mn.ff. References are cited by [rr] listed in the bibliography-reference subsection, 999, at the end of the background description and numbered by [9mn.rr] where rr is sequential and starts from 1 while 9mn is the section where the reference is first cited. Problems and solutions follow the bibliography-reference section and are numbered by P9mn.pp where pp is the problem number.

900 INTRODUCTION

The reliability or operation time-to-failure (TTF_{OP}) of multi-million-transistor integrated circuit (IC) chips is determined by the TTF of the transistors and conducting lines interconnecting the transistors. Modern IC chips guarantee a 10-year TTF_{OP}. The TTF_{OP} is defined by a specified change of some transistor characteristics beyond which the designed IC performance (power dissipation, signal distortion, error rate, ...) cannot be maintained, for examples, the transistor's leakage current, current amplification factor, output current magnitude, and transconductance. Obviously the $TTF_{OP}\geq 10yr$ requirement must be determined by some accelerated methods to assess the reliability of a new transistor technology in a short time in order to assure competitive time-to-market. The traditional voltage acceleration method uses stress voltages much higher than the circuit-operation voltage to get large acceleration or stress-time reduction, and then empirically extrapolates the short-stress-time TTF-vs-V_{STRESS} data down to the circuit-operation voltage to give the TTF_{OP}. To keep the new technology development cycle short, a three-decade extrapolation of accelerated-stress data from a 4.5-decade short-stress-time range (10s to $100hr=3.6x10^5s$) to give the 10-year TTF_{OP} ($3.2x10^8s$) is required. However, the very strong exponential voltage dependence of TTF on V_{STRESS} and the empirical extrapolation formula require that the stress voltages must be kept as low and close to the operation voltage as possible to minimize extrapolation error while still maintaining a short stress-and-measure (SAM) time on the test transistor patterns which are costly and time-consuming to fabricate (2-week nominal). The traditional voltage-acceleration scheme not only suffers from the empiricism of the extrapolation formula but also the assumption of a continuous voltage dependence of the failure rates that neglects the inherent threshold kinetic energies or applied voltages of each failure mechanism. TTF extrapolation is becoming increasingly tedious (multi-parametered) and unreliable as the transistor dimension decreases below one micron and the operation power supply voltage lowers to less than 3.3V in present and future integrated circuits. Thus, the empirical extrapolation method for TTF_{OP} must be eventually replaced by physics-based material and device models in order to provide accurate extrapolation to 10-year TTF_{OP} from a small sample of short stress-time data. This appendix provides the material-device physics backgrounds and illustration examples for predicting the TTF_{OP} from analytical solutions using measured and assumed values of fundamental material parameters. The presentation is given in four sections: interface traps, oxide traps, acceptor hydrogenation and electromigration. Fundamental and TTF-analysis problems and some solutions are given at the end of the appendix.

910 INTERFACE TRAPS – Introduction

Interface trap is an abbreviated generic name for electronic traps at the SiO_2/Si interface and other heterojunction interfaces. Interface is defined as a thin surface area between two electrically different materials. The interface contains several atomic planes of host and impurity atoms. Within the interfacial layer, the material composition and hence the electrical property are assumed constant. Thus, the concentration of an atomic imperfection species (a physical defect or chemical impurity) and carriers (electrons and holes) are the areal density or areal concentration (frequently called total density or total number) with the dimension of particle number per unit area. In the bulk of the transistor, the generation-recombination-trapping-tunneling (grtt) rates per unit volume in the volume element (dxdydz in cm^3) at a given location and time, (x,y,z,t), have the unit of number of grtt event per unit volume per second (grtt#/cm^3-s). In an interfacial layer, their dimension is (grtt#/cm^2-s). At the interface, the grt rates are often expressed as interface or surface recombination velocity (cm/s) and the less-frequently-used surface or interface generation and trapping velocities, while the tunneling rate is generally expressed as the tunneling current density (area density in Ampere/cm^2 or A/cm^2) by multiplying with the particle charge (with sign).

The simple dimensional difference between volume and interface or surface concentrations and the more complicated surface or interface recombination velocity have proven to be difficult concepts for beginners and even experienced transistor technologists and engineers due to a lack of exposure to the simple derivation mathematics of areal density from volume density, and of velocity from volume rate. The algebra is now given. Let the interfacial layer be lying between x_1 and x_2 which can be functions of the location (y,z) in the layer and even time, t, i.e., $x_1(y,z,t)$ and $x_2(y,z,t)$, if the layer is not planar and the shell's thickness is not constant in space and time (growing or thinning). This simple relationship is:

$$\int_{x_1}^{x_2} f_V(x,y,z,t)dx = \int_{x_1}^{x_2} f_S(y,z,t)\delta(x_{12})dx = f_S(y,z,t) \qquad (910.1)$$

The first integrand, $f_V(x,y,z,t)$, is a volume concentration (particle/cm^3) or volume grtt rate (grtt#/cm^3-s). The second integrand, $f_S(y,z,t)$, is the surface-interface or areal concentration (particle/cm^2) or surface-interface areal grtt rate (grtt/cm^2-s). The symbol $\delta(x_{12})$ is the Dirac delta function evaluated at some point x_{12} between x_1 and x_2 where

$x_2 > x_1$, such as the mid point of the layer thickness at a position (x,y) in the layer and at a certain time t, $x_{12}(x,y,t) = (x_1 + x_2)/2$. Thus, the decomposition, $f_V(x,y,z,t) = f_S(y,z,t)\delta(x_{12})$, is based on the simple physical assumption just made, i.e., the specific property (concentration or volume grrt rate) is spatially constant along x in the area element, dydz, (which could still be time dependent).

The recombination velocity concept is also given by (910.1) if $f_V(x,y,z,t)$ is the per trap-concentration or unit volume grtt rate defined by the product of (i) the grtt volume capture rate at the trap, c(event-cm^3/s), and (ii) the volume concentration of the carriers n(x,y,z,t) and traps $n_T(x,y,z,t)$ (particle/cm^3), and divided by the trap concentration $N_{TT}(x,y,z)$ (particle/cm^3). Thus, it has the dimension of velocity as shown as follows.

$$f_V(x,y,z,t)dx = [(\ c\ )\cdot(\ n\ )\cdot(\ n_t\ )/(N_{TT}\ )]\cdot(dx)$$
$$\Rightarrow [(cm^3/s)\cdot(cm^{-3})\cdot(cm^{-3})/(cm^{-3})]\cdot(cm) = cm/s \qquad (910.2)$$

If the interfaces are inside the electrically active region of a transistor, or within a diffusion length of the active layer or volume, then electron-hole recombination-generation at the traps in these interfacial layers will affect the transistor's electrical characteristics and circuit performance. When the interface trap density changes due to electrical stresses (high voltage, high electric field, and high current density) or high temperatures during the operation of the transistor, the transistor will degrade and the IC will eventually fail to perform the designed electrical function (switching amplification, and specific digital and analog signal processing).

The bare surface is a solid/gas or solid/vacuum interface. It is of fundamental importance to provide the baseline for understanding the solid/solid interfaces in transistors and integrated circuits. The practically important solid/solid interfaces are composed of any pair of materials from the following list which also gives the symbols we shall use: crystalline Si (cSi or just Si), SiO_2 (SiO_2 grown on crystalline Si), polycrystalline Si (poly-Si or just poly), SiO_2 grown on poly-Si (polySiO_2), amorphous Si (aSi), SiO_2 grown on amorphous Si (aSiO_2), and crystalline SiO_2 (cSiO_2); metals (Al, Au, Cu with generic symbol M), metal-oxides (Al_2O_3, CuO, Cu_xO_y with the generic symbol M_xO_y); refractory metals (W, Ti, Ta, Mo, with generic symbol Rm but not using R to avoid conflict with Radicals), refractory metal oxides (Rm_xO_y), silicides or refractory metal silicides (Si_xM_y or Si_xRm_y), and other semiconductors (mainly Ge or Ge_xSi_{1-x}). Using our junction symbol with the interface represented by the slash '/', some examples are: poly-Si/Si or poly/Si, aSi/Si, SiO_2/Si, and $cSiO_2$/Si for the ultimate silicon integrated circuits. The bare, clean and ordered silicon surface will be described first which is given in the next section.

911 Surface States on Clean Silicon Surfaces

A review of the interface traps at the SiO_2/Si interface was given [1-4] which included also electronic traps due to the dangling silicon bonds on bare or clean and atomically flat silicon surfaces (Si bar cleaved in ultrahigh vacuum chamber) not covered by spotty oxide films or residual impurities [1]. This bare surface was of fundamental interest to the semiconductor physicists in the 1950's (Bardeen and Brattain obtained the clean surface by cleaving Si and Ge in vacuum and cleaning the surface in vacuum via Argon ion bombardment and heating). This interest has continued to this day, using these same surface cleaning techniques, but in Ultra High Vacuum (UHV) chambers, in order to study the fundamentals of molecular beam epitaxial growth. The 1950's data obtained on bare Si surfaces can still serve as a baseline for today's technologically important surface, the oxidized silicon, since the residual intrinsic interface traps on oxidized crystalline silicon in MOSTs and BJTs due to silicon dangling and oxygen bonds may be similar to the intrinsic (Si dangling bonds) and impurity (oxygen dangling bonds) surface states observed on the bare silicon surfaces in vacuum.

Table 911.1
Atomic and Cut-Bond Densities of Un-reconstructed Si Surfaces

FACE	2-D Cell Shape	One Cell Has Area (a^2)	Atom	Cut-Bond	Areal Density ($10^{14}cm^{-2}$) Atom	Cut-Bond	Comments
*****	*******	****	****	****	*****	******	***********************
(100)	square	1	2	4	6.780	13.56	Peroxide bond Si-O-O-Si
(100)	square	1	2	2	6.700	6.780	Monoxide bond Si-O-Si
(110)	rectangle	$\sqrt{2}$	4	4	9.589	9.589	Exclude bonds in plane
(110)	rectangle	$\sqrt{2}$	4	8	9.589	19.178	Include bonds in plane
(111)	diamond	$\sqrt{(3/4)}$	1	1	7.832	7.832	Face A
(111)	diamond	$\sqrt{(3/4)}$	1	3	7.832	23.498	Face B
(111)	diamond	$\sqrt{(3/4)}$	1	2	7.832	15.665	Average (A+B)/2

a=5.430A (Si). At 300K. From [1].

The Si dangling bond density of un-reconstructed bare silicon surface was tabulated by earlier authors that contained some errors. The corrected values computed by this author is given in Table 911.1.

Allen and Gobeli [1] measured the surface trap density on clean (111) Si and fitted the measured surface charge density to

$$Q_{ST}(U_S \equiv qV_S/k_{BT}) = -\ 2.44\times10^{12}\sinh(U_S+9.7)\ q/cm^2 \qquad (911.1)$$

with a width of about $2\times6.45k_BT \simeq 12.9k_BT \simeq 333$meV at 300K. Their data can also be fitted to two peaked DOS (electronic Density of States) separated by about 0.3eV each with a density approximately equal to the silicon dangling bond density on the (111) face, $7.8\times10^{14}cm^{-2}$. The normalized surface potential, $U_S=qV_S/k_{BT}=(E_F-E_I)/k_BT$, is the Fermi level at the surface measured from the intrinsic Fermi level at the surface. This experimental charge distribution gives a neutral Fermi level position at $E_{F\text{-neutral}} = E_I - 9.7k_BT = E_I - 220$meV or 220meV below the silicon intrinsic Fermi level, E_I, which is located near the Si midgap, E_{MG}. The neutral Fermi level concept was first defined and used by Bardeen in 1948 to explain the independence of the metal/semiconductor Schottky barrier height on the metal type. (See P911.2 for an alternative and more realistic DOS extracted from the DOS of the clean, bare, and unconstructed Si surface.)

912 Interface Traps on Oxidized Silicon

From a variety of experiments over the last three decades using MOS diode and transistor structures [1-4], three interfacial traps in the DOS spectra have been repeatedly observed by many researchers at the SiO_2/Si interfaces of oxidized crystalline silicon [1-4]. These were attributed by us to three intrinsic interfacial defects rather than impurities due to the wide variety of sample preparation conditions containing different impurities and the different generation and annealing rates, although the absence of an impurity-interface trap in the experimental data has been puzzling but experiments designed to seek their presence have not been undertaken systematically. Two have peaked DOS spectra with broadened FWHM (Full Width at Half Maximum). The peaks are located at approximately 0.35eV above and below the Si midgap. The third has a broad featureless U-shaped DOS which extends over the entire Si energy gap and its magnitude rises rapidly near the Si band edges, $E_{V\text{-}Si}$ and $E_{C\text{-}Si}$.

From simple atomic physics or elementary properties of the solutions of the Schrödinger equation, a peaked DOS must originate from a random or regular array of point defects in the volume or surface being measured. A DOS peak is the signature of a defect. The condition for the presence of a DOS peak is that the defect must give a sufficiently localized and large perturbation to the crystalline periodic potential in order that the Schrödinger equation has a bound solution. To give a measurably large DOS amplitude, there must also be a sufficiently large number of the defect distributed over the measured volume or surface area. Thus, the physical size of each point defect is small (i.e. localization, consisting of one to several but not many impurity and/or dislocated host atoms) and the atomic environment of each defect must be nearly identical over the volume or surface being measured (i.e. to give the narrow peak). The two DOS peaks are not present simultaneously in all experiments, thus they must be associated with two specific and distinct intrinsic defects. Different generation and annealing kinetics further indicated that these are two different defects rather than the multi-electron or amphoteric (one-electron and one-hole) bound states of the same defect species. These properties and some recent MOST measurements, made at higher sensitivity and on samples with tighter fabrication control from oxides grown in the production furnaces of the very-high-yield multi-million transistor microprocessor chips, have suggested that the two interfacial DOS peaks are related to the silicon and oxygen dangling bond defects, $Si_3{\equiv}Si\bullet$ at $E_{MG} - 0.4eV$ (tentatively identified as a donor due to negative gate voltage shift) and $(Si\text{-}O)_3{\equiv}Si\text{-}O\bullet$ at $E_{MG} + 0.3eV$ (identified unequivocally as a donor).

An alternative model is the four silicon dangling bonds of different atomic environment represented by the chemical formula $(Si_xO_{3-x}){\equiv}Si\bullet$ where x= 0,1,2,3 corresponding to (x=3) $Si_3{\equiv}Si\bullet$, (x=2) $(Si_2O){\equiv}Si\bullet$, (x=1) $(SiO_2){\equiv}Si\bullet$, and (x=0) $O_3{\equiv}Si\bullet$ which is identical to $(SiO)_3{\equiv}Si\bullet$. The x=3 configuration, $Si_3{\equiv}Si\bullet$, is known as the **trivalent silicon** and has the silicon dangling bond pointing towards the SiO_2/Si interface from the Si substrate into a cavity where the bridging oxygen is missing i.e. the bridging oxygen vacancy. This was first detected in 1971 by Nishi from EPR spectra and designated as the P_b center [2]. The cavity atomic configuration was later resolved by detailed analysis of the EPR spectra and labeled the P_{b0} center on the (111) surface while a second center was labeled P_{b1} on the (100) surface. The x=0 configuration, $(SiO)_3{\equiv}Si\bullet$, is the trivalent silicon dangling bond in the oxide pointing towards the SiO_2/Si interface.

The broad FWHM reflects two facts: (1) the basic DOS measurement limitation with a resolution of about $4k_BT$ or 100meV, and (2) the random environment around each defect center causing random variation of the amplitude and range of the perturbation potential from each defect due to (2a) random bond length and angle of the interfacial Si-Si and Si-O bonds and (2b) the random x-distribution of $(Si_xO_{3-x})\equiv Si\bullet$ surrounding the dangling bonds distributed over the entire interface area of the MOST. Zeeman resonance in EPR of the P_b center on (111) face of Si analyzed by Bower in 1983-1986 [2] show a random variation of bond angle of 0.5^o and bond length of 0.02A and the SiO_2/Si interface is atomically flat over distances from 15A to 300A which is interrupted by steps of one atomic plane high.

The characteristics of the solutions of the Schrödinger equation also suggest that the third, U-shaped, DOS could come from the random distribution of the properties of the point defects: defect density or number in the elemental volume or surface area being measured, and strength of the perturbation. The U-shape suggests a defect number distribution, with a larger number of defects (high DOS) of small perturbation strength (small energy shift or band edge states) and a smaller number of defects (low DOS) of larger perturbations (energy shift to the Si midgap).

There is also a nearly constant DOS in the Si midgap range whose density is low and not affected by ambient and various applied stress except heating and cooling rates and durations. This suggests that the perturbation is the random distribution of the Si-Si and Si-O bond lengths and angles at the amorphous-SiO_2/crystalline-Si interface which should be uniform in size and number distribution. Indeed experimental confirmations have been abundant [4,5,6] showing higher density of the U-shaped DOS at higher post-oxidation heating temperatures in a nonoxidizing ambient, such as prolonged heating of a 1000C thermally grown oxide in vacuum or dry argon at 1200C.

There is a transition layer from the crystalline Si lattice to the amorphous SiO_2 lattice with the composition SiO_x with x=0 at the Si-side of the Si/SiO_2 transition layer and x=4 at the SiO_2-side of the Si/SiO_2 transition layer. These are called the suboxides. The thickness and presence of this transition layer should be highly dependent on the oxidation conditions, such as the oxidation ambient and heating-cooling rates. For example, post oxidation 'anneal' (POA) in vacuum at a higher temperature than the oxidation temperature should greatly increase the thickness of this oxygen deficient transition layer and give a corresponding increase of the interfacial traps. The increase of the interface trap density by this POA step has been observed for many decades, including a threshold voltage shift to 150V in MOST with 2000A gate oxide heated in dry argon at 1200C fabricated by one of my former graduate students in the mid-1970's. Measurement of the suboxide abundance in the interfacial layer by the X-Ray Photoemission Spectroscopy (XPS) was reported by Grunthaner in 1987 [3]. A comparison of the SiO_x abundance at several oxidation conditions with the silicon dangling bond density was made which is given in Table 912.1. In this table, a comparison with the dangling bond density on ideal bare Si surface, listed in Table 911.1, is also given.

Table 912.1
Process Dependence of the Suboxide Density and the SiO_2/Si Interface

FACE	Oxidation Condition	Density ($10^{14}cm^{-2}$) Atom	CutBond	Si^{+}	Si^{+2}	Si^{+3}	Total	Monolayer Covered %
*****	*****************	****	*******	*****	*****	*****	*****	*********
(100)	theory (peroxide)	6.78	13.56	0.	13.56	0.	13.56	100
	theory (monoxide)	6.78	6.78	0.	6.78	0.	6.78	100
	900C, dry			0.2	2.7	3.5	6.4	94
	900C, wet			0.1	3.0	2.7	5.8	85
	900C, wet POA			0.5	2.1	2.5	5.1	75
	1000C, dry			0.	3.2	2.7	5.9	87
	1000C, dry POA			0.5	2.3	2.6	5.4	79
(111)	theory (face-A)	7.83	7.83	7.83	0	0	7.83	100
	theory (face-B)	7.83	24.498	0	0	7.83	7.83	100
	theory (A+B)/2	7.83	15.66	7.83	0	7.83	15.66	100
	900C, dry			3.6	0.2	2.7	6.5	83

From reference [3].

The notation listed in the above table is the classical chemical notation defined as follows in contrast to the solid-state notation of Si^{+4} for the silicon atomic core without the four valence electrons. The monovalent Si^{+} is a silicon atom bonded by one O and three bulk Si's which are bonded to the Si's in the bulk, and represented by the chemical formula $Si_3{\equiv}Si_3{\equiv}Si\text{-}O\text{-}Si{\equiv}(O\text{-}Si)_3$. The divalent silicon, Si^{+2}, is $Si_2{=}Si{=}(O\text{-}Si)_2$. The trivalent silicon, Si^{+3}, is $Si_3{\equiv}Si\text{-}Si{\equiv}(O\text{-}Si)_3$. For reference, the covalent Si in the silicon bulk is Si^0 or $Si_2{=}Si{=}Si_2$ or Si_4::::Si, and in the oxide film, the 'fully ionized' tetravalent silicon, Si^{+4}, is $(Si\text{-}O)_2{=}Si{=}(O\text{-}Si)_2$.

The correlation with the oxidation condition just anticipated is observed in Table 912.1 which shows that the monolayer coverage has decreased significantly from 87% coverage to 79% when the 1000C dry oxide grown in the (100) face of Si is annealed after oxidation in a dry inert ambient at 1000C. The largest increase is the monovalent Si^{+} from $0.0\times10^{14}cm^{-2}$ to $0.5\times10^{14}cm^{-2}$. One would expect a connection between the interface trap densities and the concentration of the suboxide species, $(Si_{4-x}O_{x})$::::Si, because the suboxide bonds may be ruptured to give dangling bonds which are interface traps. The partially bonded species may have eight single-dangling bond configurations given by $Si_3O_y\equiv Si\bullet$ and $Si_3O_y\equiv SiO\bullet$ where y=0,1,2 or 3; eighteen double-dangling bond configurations; and eight triple-dangling-bond configurations. The four-dangling-bond configuration would be an interstitial. If the Si bond is ruptured, the single Si dangling bond or the trivalent silicon, $Si_3\equiv Si\bullet$, would have the largest concentration in the (111) plane, $7.83\times10^{14}cm^{-2}$ as indicated in Table 911.1 for the ideal bare Si surface. This is consistent with the largest Si^{+} concentration measured on the 900C dry oxide grown on the (111) surface, $3.6\times10^{14}cm^{-2}$.

The last column of the above table, Table 912.1, shows that 900C-dry (100) oxide has 94% of the Si at the SiO_2/Si bonded by oxygen while the 900C-dry (111) oxide has only 83% of the Si bonded by oxygen, given an experimental ratio of 0.94/0.83=1.13. In comparison, the theoretical ratio of the ideal (100) to (111) ratio is 15.66/13.56=1.15 using the (A+B)/2 two face average for the thermally oxidized (111) interface. Thus, the correlation of the Si• dangling bond density between the ideal bare surface and the SiO_2/Si interface of this specially grown oxide is not only qualitative but also quantitative. (See P912.3 for further considerations and applications of this correlation.)

913 Interface Trap Creation-Destruction by Hydrogen

The important residual interface traps are probably the two intrinsic defects, the silicon dangling bond, $(SiO)_3\equiv Si\bullet$, and the oxygen dangling bond, $(SiO)_3\equiv SiO\bullet$. Some of these dangling bonds may be inactivated, such as by anneal at 450C in forming gas, $10\%H_2+90\%N_2$, and by unintentional hydrogenation during fabrication. Other less well characterized interface traps are those electron-hole binding centers containing the impurities, N and F, which are intentionally added during oxidation to give higher bond strength than H, and those from the unavoidable dopant impurities (B, In, P, As) employed in the current and future silicon ULSI MOSTs and BJTs. These interfacial traps increase the p/n junction leakage current and the standby power dissipation, lower the BJT current amplification factor, increase low-frequency 1/f noise, and cause instability or lower the TTF_{OP} when the bridging hydrogen is released via breaking the electron bonds by energetic electron impact or thermal hole capture. We will focus on the fundamental failure mechanisms and kinetics from the hydrogenation and dehydrogenation of the silicon and oxygen dangling bonds.

The interface traps due to silicon and oxygen dangling bonds can be deactivated electrically by hydrogenation. The deactivated dangling bonds can be reactivated by dehydrogenation. Change of their electrical activities would change the areal or surface charge density at the SiO_2/Si interface and also the electron-hole generation -recombination-trapping rates, both would degrade and destablize the MOS and bipolar transistor characteristics covered by the oxide. There are five known technologically important elementary dehydrogenation processes which create silicon and oxygen dangling bonds and which have been observed in experiments and whose kinetics have been analyzed. These are listed in (913.1A) to (913.5B).

The first four elementary processes given by (913.1A) to (913.4B) involve the fundamental process of removing one electron from the 2-electron hydrogen-terminator bond in Si:H and SiO:H, and the removal energy then releases the second electron to cause the unbonded hydrogen ion or proton to move away. If the electric field is low, the proton can capture an electron to become a hydrogen atom and continue to move away. In high electric field region with low or zero electron concentration, the ionized hydrogen or proton will move away without capturing an electron. The fate of the hydrogen could be the capture at a hydrogen trap or recombination to form H_2 which would evaporate if at the surface. For expediency, these hydrogen migration-reaction processes while moving away from the bond site are lumped into the symbol $H{\bullet}\uparrow$. The agent that removes the first bond electron can be heat or phonon, photon($E_{H\text{-}Bond} \simeq 3eV < h\nu < E_{G\text{-}SiO2} \simeq 8.5eV$), oxide hole capture, energetic electron impact ($E_{H\text{-}Bond} \simeq 3eV < KE_{electron} < E_{G\text{-}SiO2} \simeq 8.5eV$), and hydrogen reaction.

The fifth process, (913.5A) and (913.5B), is the chemical reaction with an atomic hydrogen to 'reduce' the hydrogenated silicon and oxygen. This reaction leaves a Si and O dangling bond and releases the hydrogen. It is thought to be an important mechanism of interface trap creation in oxide-protected p/n junctions such as the emitter/base of BJTs and the drain/body junction of MOSTs.

$$\mathrm{Si{:}H + heat \rightarrow Si\bullet + H\bullet\uparrow} \quad (913.1A)$$
$$\mathrm{SiO{:}H + heat \rightarrow SiO\bullet + H\bullet\uparrow} \quad (913.1B)$$

$$\mathrm{Si{:}H} + h\nu \rightarrow \mathrm{Si\bullet + H\bullet\uparrow} \quad (913.2A)$$
$$\mathrm{SiO{:}H} + h\nu \rightarrow \mathrm{SiO\bullet + H\bullet\uparrow} \quad (913.2B)$$

$$\mathrm{Si{:}H} + h^+ + e^- \rightarrow \mathrm{Si\bullet} + H^+ + e^- \rightarrow \mathrm{Si\bullet + H\bullet\uparrow} \quad (913.3A)$$
$$\mathrm{SiO{:}H} + h^+ + e^- \rightarrow \mathrm{SiO\bullet} + H^+ + e^- \rightarrow \mathrm{SiO\bullet + H\bullet\uparrow} \quad (913.3B)$$

$$\mathrm{Si{:}H} + h^{+*} \rightarrow \mathrm{Si\bullet + H\bullet\uparrow} \quad (913.4A)$$
$$\mathrm{SiO{:}H} + h^{+*} \rightarrow \mathrm{SiO\bullet + H\bullet\uparrow} \quad (913.4B)$$

$$\mathrm{Si{:}H} + e^{-*} \rightarrow \mathrm{Si\bullet + H\bullet} + e^- \rightarrow \mathrm{Si\bullet + H\bullet\uparrow} + e^- \quad (913.5A)$$
$$\mathrm{SiO{:}H} + e^{-*} \rightarrow \mathrm{SiO\bullet + H\bullet} + e^- \rightarrow \mathrm{SiO\bullet + H\bullet\uparrow} + e^- \quad (913.5B)$$

$$\mathrm{Si{:}H + H\bullet \rightarrow Si\bullet + 2H\bullet\uparrow} \quad (913.6A)$$
$$\mathrm{SiO{:}H + H\bullet \rightarrow SiO\bullet + 2H\bullet\uparrow} \quad (913.6B)$$

There are a variety of conditions encountered in applications where combinations of several of these elemental processes will occur. The following are four examples of several simultaneous or parallel pathways each with several serial pathways. The symbol $\mathrm{H\bullet\uparrow}$ implies all hydrogen removal pathways and n(eh) is the electron and hole plasmons created by the keV electron beam or x-Ray.

High Current Density in BJT and MOST ($\sim 10^6$A/cm^2)

$$\mathrm{Si{:}H} + 2e^- + h^+ \rightarrow \mathrm{Si{:}H} + e^{-*} \rightarrow \mathrm{Si\bullet + H\bullet\uparrow} \quad (913.7A)$$
$$\mathrm{SiO{:}H} + 2e^- + h^+ \rightarrow \mathrm{SiO{:}H} + e^{-*} \rightarrow \mathrm{SiO\bullet + H\bullet\uparrow} \quad (913.7B)$$

Electron Plasmon in High Voltage BJT-MOST

$$\mathrm{Si{:}H} + e^{-**} \rightarrow \mathrm{Si{:}H} + 2e^{-*} + h^+ \rightarrow \mathrm{Si\bullet + H\bullet\uparrow} + e^- \quad (913.8A)$$
$$\mathrm{SiO{:}H} + e^{-**} \rightarrow \mathrm{SiO{:}H} + 2e^{-*} + h^+ \rightarrow \mathrm{SiO\bullet + H\bullet\uparrow} + e^- \quad (913.8B)$$

keV Electron Beam Lithography

$$\mathrm{Si{:}H} + e^{-***} \rightarrow \mathrm{Si{:}H} + e^- + h^+ + e^{-**} \rightarrow \mathrm{Si\bullet + H\bullet\uparrow} + e^{-**} \quad (913.9A)$$
$$\mathrm{SiO{:}H} + e^{-***} \rightarrow \mathrm{SiO\bullet H} + e^- + h^+ + e^{-**} \rightarrow \mathrm{Si\bullet + H\bullet\uparrow} + e^{-**} \quad (913.9B)$$

x-Ray Lithography

$$\mathrm{Si{:}H} + h\nu^{***} \rightarrow \mathrm{Si{:}H} + e^- + h^+ + n(eh)^{**} \rightarrow \mathrm{Si\bullet + H\bullet\uparrow} + n(eh)^{**} \quad (913.10A)$$
$$\mathrm{SiO{:}H} + h\nu^{***} \rightarrow \mathrm{SiO\bullet H} + e^- + h^+ + n(eh)^{**} \rightarrow \mathrm{Si\bullet + H\bullet\uparrow} + n(eh)^{**} \quad (913.10B)$$

The interface traps can be electrically deactivated by hydrogenation, oxidation, and thermal anneal which are represented by the following reaction equations. It is evident that hydrogenation given by (913.11A) and (913.11B) are the most likely processes since the isolated defect concentration is much higher than the concentration of the complex consisting of two defects.

```
Si•  + H•                  →  Si-H        (913.11A)
SiO• + H•                  →  SiO-H       (913.11B)
Si•  + Si•  + heat         →  Si-Si       (913.12A)
Si•  + Si•  + O: + heat    →  Si-O-Si     (913.12B)
Si•  + SiO• + heat         →  Si-O-Si     (913.12C)
```

914 Interface Trap Generation by VUV Light

Multiple interfacial DOS peaks were observed after MOS capacitors were exposed to 10.2eV Vacuum Ultra-Violet (VUV) light from a hydrogen discharge lamp under positive gate voltage [5,6]. The VUV light generated holes at the surface of the SiO_2 are injected into the oxide by the positive gate voltage and then captured by hydrogenated interface traps. High densities of these traps appeared in oxides with post oxidation anneal (POA) in oxygen and hence are probably oxygen dangling bonds of the interfacial suboxides from incomplete oxidation at the low POA temperature in oxygen.

The DOS plots are shown in Fig.914.1 (a), (b) and (c), and the relaxation kinetics in (d). There are five DOS components. Four are increasing with VUV light exposure time. They are the three peaks in the upper half of the Si energy gap, labeled P1, P2 and P3, and were attributed to the oxygen dangling bond. The fourth is the relatively flat midgap states, labeled M, which were attributed to the wide random distribution of the various silicon dangling bonds. Their growth in concentration was attributed to dehydrogenation to be illustrated below. The fifth component is relatively insensitive to the VUV exposure with a nearly constant U-shaped DOS, which is labeled U_0 and was attributed to the band-tail states from the random distribution of the Si-Si and Si-O bond angles and lengths. Experimental separation of the M and U_0 components is difficult due to similar U-shape DOS. The different growth and annealing rates shown in Figs.914.1(a) and (c) indicate their different origins. The VUV generation parameters of the five interface traps are listed in Table 914.1.

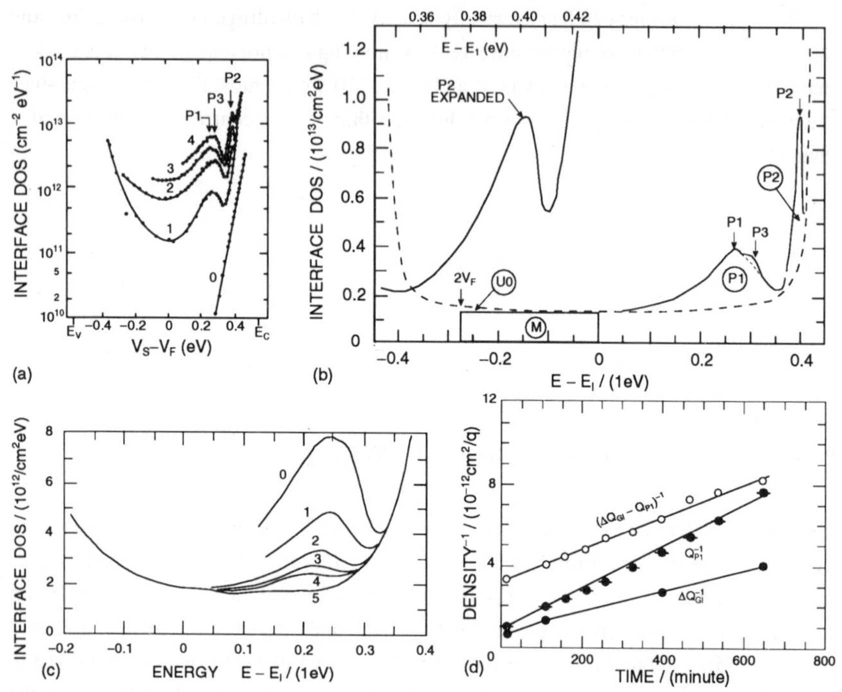

Fig.914.1 The interface DOS in p-Si MOS capacitor generated by oxide hole capture during exposure to 10.2eV photons under positive gate voltage and during thermal anneal at room temperature. **(a)** Original data during exposure. Curve 0: pre-exposure. **(b)** Enlarged DOS after exposure 4. **(c)** Relaxation: thermal anneal at room temperature. **(d)** Relaxation kinetics. From [2,4,6].

Table 914.1
10.2eV Photo-Generation Parameters of Five Interface Traps

TRAP LABEL	EFFICIENCY (10^{-3} trap/hole)	Cross-Section ($10^{-16}cm^2$)	Steady-State Density ($10^{11}cm^{-2}$)
U0	0	0	0.45
M	4.3	80	5.2
P1	6.1	140	4.2
P2	1.4	23	4.0
P3	0.3	-	-

From [6].

The three DOS peaks in the upper energy gap of Si (P1, P2, P3) are donor-like, i.e. with positive and neutral charge states, which were deduced from the observed positive gate voltage shift during thermal anneal. The three peaks grew simultaneously during VUV-PHI (Photo Hole Injection) stress and relaxed also simultaneously during thermal anneal at room temperature. The relaxation or decay followed the second order kinetics with the $1/t_{anneal}$ dependence shown in Fig.914.1(d) which led this author to consider two models: (i) hydrogenation of the oxygen dangling bond on the four suboxides and (ii) multi-electron trap [6]. Their underlying physics are now described to serve as a background for additional classroom (and also research) problems and solutions.

Consider first the multi-electron bound states from the oxygen dangling bond trap on the silicon side of the SiO_2/Si interface, $Si_3{\equiv}Si\text{-}O\bullet$. The binding energies may be estimated from the experimental energy levels of electrons bound to an oxygen in vacuum, using the dielectric approximation to take into account the valence electron screening of the oxygen perturbation potential at the SiO_2/Si interface. The electron affinity and the first three ionization potentials in vacuum from the Handbook of Physics and Chemistry are given in Table 914.2.

Table 914.2
Selected List of Electron Affinity, Ionization Potential and Binding Energy of Electrons in SiO_2, Si-Crystal and in Si and Impurity Atoms

(all in eV unit)						--SiO_2/Si--		-- SiO_2 --	
ELEMENT	SiO_2	Si(c)	Si	O	TRAP	Si•	O•	Si•	O•
*******	****	*****	******	******	*********	*****	*****	*****	*****
EA	0.9	4.018	1.427	1.466	acceptor	0.023	0.024	0.094	0.096
1st IP	9.4	5.188	8.151	13.618	1st donor	0.134	0.224	0.536	0.895
2nd IP			16.345	35.116	2nd donor	0.268	0.577	1.074	2.308
3rd IP			33.492	54.934	3rd donor	0.558	0.903	2.198	3.612
4th IP			45.141	77.412	4th donor	0.742	1.272	2.968	5.090

EA Electron Affinity in vacuum.
IP Ionization Potential in vacuum. E_G=8.5eV(SiO_2), 1.17eV(Si).
(c) crystalline Si.

Let us use oxygen's electron affinity in vacuum, 1.466eV, to estimate the first acceptor level of oxygen at the SiO_2/Si interface. Since the binding energy is inversely proportional to the square of the dielectric constant (as indicated by the Bohr formula for hydrogen although that is for an attractive Coulomb potential while the electron affinity is the binding energy to a neutral atom), we take the average value of the static dielectric constant of SiO_2 and Si for the SiO_2/Si interface, i.e. (3.9+11.7)/2=7.8. Then, the binding energy of an electron to the neutral oxygen at the SiO_2/Si interface would be $1.466\text{eV}/7.8^2 = 0.024095\text{eV}$. This is too shallow a level or too close to the conduction band edge of Si to be observed by the MOS HFCV method, however, the static dielectric screening is credible because of the large bond electron orbit which enclosed many valence electrons in the SiO_2 and Si lattices, although the dielectric screening based on the Coulomb potential overestimated the short-range potential of the neutral oxygen. But, we are looking for a theoretical estimate of the electron binding energy for the donor-like DOS peak at $E_{C\text{-}Si} - 0.3 \pm 0.1$ eV. Thus, we use the same procedure to estimate the oxygen donor levels by dielectrically screening the ionization potentials of the oxygen atom in vacuum. The screening using the Bohr formula is now correct since the first electron is bound to a singly positively charged oxygen core. This gives the values in the O column under SiO_2/Si in Table 914.1. The SiO_2/Si dielectrically screened first ionization potential is 0.224eV which is close to the observed binding energy or DOS peak at $E_{C\text{-}Si} - 0.3 \pm 0.1$ eV. This binding energy is not large, hence the bound orbit is not small and contains many SiO_2 and Si valence electrons to make the static dielectric screening credible. The second and third donor levels are deeper and hence have smaller bound electron orbits which would make the assumed full static dielectric screening less effective and which in turn would increase the estimated binding energies to values larger than those listed in Table 914.1. This would place the second and higher donor energy levels inside the Si valence band so they are always filled by electrons and cannot trap electrons or holes. Similarly, we can estimate the interface DOS peak from a dangling Si• bond which gives $E_{C\text{-}Si} - E_{Dn}$ (n=2,3,4) = 0.134eV, 0.268eV, 0.558eV and 0.742eV, however, a distinct DOS peak attributable to the Si• dangling bond has not been established by systematic experiments, such as using a vacuum heated oxide to generate a high concentration of oxygen vacancies or trivalent silicon interface traps. Since the second and third donor levels are deeper, the multi-electron oxygen model cannot account for the three DOS peaks.

An alternative possible origin of the multiple DOS peaks is several oxygen dangling bond configurations on the suboxides, $(Si_{3-y}O_y)\equiv SiO\bullet$ where y=0,1,2,3, i.e. P1 is $(Si_3)\equiv Si{-}O\bullet$, P2 is $(Si_2O)\equiv Si{-}O\bullet$, and P3 is $(SiO_2)\equiv Si{-}O\bullet$. The interface trap generation reactions are listed below. There are six atomic configurations on the (100) interface for the four hydrogen bond breaking reactions each with a different O-H bond strength due to the different number of adjacent bond-center oxygen atoms which also gives different generation rates. The least amount of hole energy is required to break the

O-H bond in $(Si_3){\equiv}SiO\text{-}H$ because $(Si^{+4})_3{\equiv}Si^{+4}\text{–}O\bullet$ is less coulombic attractive to the hydrogen atom, H•, than $[(Si^{+4})_2O]{\equiv}Si^{+4}\text{–}O\bullet$ and $[(Si^{+4})O_2]{\equiv}Si^{+4}\text{–}O\bullet$. Thus, P1 > P2 > P3 in hydrogen binding energy and generation cross-section while the reverse is true for the electron binding energy or $E_C - E_T$. These are consistent with Table 914.1.

Reaction	Products	O-H BOND STRENGTH	Electron TrapEngy		
$(Si_3){\equiv}SiO{:}H + h\nu \rightarrow$	$(Si_3){\equiv}SiO{:}H + h^+ + e^-$				
$\rightarrow$	$(Si_3){\equiv}SiO\bullet + H\bullet\uparrow$	Lowest	Highest	P1	(914.1)
$(Si_2O){\equiv}SiO{:}H + h\nu \rightarrow$	$(Si_2O){\equiv}SiO{:}H + h^+ + e^-$				
$\rightarrow$	$(Si_2O){\equiv}SiO\bullet + H\bullet\uparrow$		Lower	P2	(914.2)
$(SiO_2){\equiv}SiO{:}H + h\nu \rightarrow$	$(SiO_2){\equiv}SiO{:}H + h^+ + e^-$				
$\rightarrow$	$(SiO_2){\equiv}SiO\bullet + H\bullet\uparrow$			P3	(914.3)
$(O_3){\equiv}SiO{:}H + h\nu \rightarrow$	$(O_3){\equiv}SiO{:}H + h^+ + e^-$				
$\rightarrow$	$(O_3){\equiv}SiO\bullet + H\bullet\uparrow$	Highest	Lowest		(914.4)

The details of the generation kinetics of the interface traps stressed by VUV light are described by the following sixteen reaction-transport equations.

(1) Electrons and holes are generated in pairs in SiO_2 across its energy gap (E_G=8.5eV) by the 10.2eV VUV light at a thin surface layer of the oxide covered by a transparent electrode (50Å gold, Au).

$$SiO_2 + h\nu \rightarrow SiO_2 + e^- + h^+ \qquad (914.5.1)$$

(2) The photogenerated oxide electrons and holes are separated by the oxide electric field from the applied gate voltage. The oxide electrons are collected by the positive gate electrode. The oxide holes are driven into the oxide, drift and diffuse through the oxide and reach the SiO_2/Si interface. (The diffusion-drift transport equation is represented by ⇒.)

$$h^+(gate/SiO_2) \Rightarrow h^+(SiO_2/Si) \qquad (914.5.2)$$

(3) One of the two bond electrons in a hydrogenated oxygen dangling bond recombines with a hole.

$$SiO{:}H + h^+ \rightarrow (SiO\bullet H)^+ + RE \qquad (914.5.3)$$

(4) The recombination energy (designated by RE) ejects the second electron in the 2-electron O:H bond and frees the proton H^+ from the oxygen protonic trap

$$(SiO\bullet H)^+ + RE \rightarrow (SiO)^+ + H^+ + e^{-*}(\sim 2.22eV) \qquad (914.5.4)$$

The recombination energy may be estimated from the thermal activation energy of dehydrogenation of O:H or the O-H bond energy in vacuum which is 4.44eV. The O-H bond energy at the SiO_2/Si interface is smaller than that at vacuum due to the adjacent Si and O atoms. This bond energy is shared by the two bond electrons which must be supplied by thermal energy to release the proton from the oxygen protonic trap during thermal dehydrogenation. Thus the kinetic energy of the ejected energetic electron is half of the bond energy or thermal activation energy, i.e., 4.44eV/2=2.22eV. The electron is on the Si side of the SiO_2/Si interface since $(SiO)_3{\equiv}SiO{:}H$ points towards Si from SiO_2, thus the 2.22eV kinetic energy is insufficient for the electron to climb over the 3.13eV SiO_2/Si barrier to be collected by the positive gate, so it will reside in the Si surface layer at the SiO_2/Si interface.

(5) However, the 2.22eV electron will generate slightly less than two electron-hole pairs in Si because the threshold energy for electron impact generation of an e-h pair, $E^n_{pn} \simeq 1.17eV$, is slight more than the Si energy gap, 1.12eV.

$$e^{-*} + Si{:}Si \rightarrow 3e^- + 2h^+ \Rightarrow \qquad (914.5.5)$$

(6) The electron can be captured by the positively charged oxygen dangling bond which is an electron trap at the interface.

$$(SiO)^+ + e^- \rightarrow SiO\bullet \qquad (914.5.6)$$

(7) The hole drifts and diffuses into the interior of the Si substrate.

$$h^+(SiO_2) \Rightarrow h^+(\text{Si-substrate/metal contact}) \qquad (914.5.7)$$

There are two electric field-dependent pathways for H^+.

(8) Low electric field.

The proton, H^+, is neutralized or becomes a hydrogen atom H• by capturing an electron among the electrons accumulated in the silicon surface layer at the SiO_2/Si interface from both the positive gate voltage and the e-h generation by the 2.22eV electron given by (914.5.4).

$$H^+ + e^- \rightarrow H\bullet \qquad (914.5.8)$$

(9) The neutral hydrogen then diffuses away from the SiO_2/Si interface into the interior of Si and towards the gate through the SiO_2.

$$2H\bullet(SiO_2/Si) \Rightarrow H\bullet(\text{Si bulk}) + H\bullet(\text{gate}/SiO_2) \qquad (914.5.9)$$

(10) High electric field.

The proton, H^+, drifts through the Si surface space-charge layer and then diffuses in the quasi-neutral Si towards the back contact of the silicon substrate.

$$H{+}(SiO_2/Si) \Rightarrow H^+(\text{QN-Si}) \qquad (914.5.10)$$

(11) In the quasi-neutral region, the proton can capture an electron if there are electrons nearby, such as in a n-type Si.

$$H^+ + e^- \rightarrow H\bullet \quad (914.5.11)$$

(12) The neutral hydrogen then continues to diffuse towards the back contact to the Si substrate.

$$H\bullet(QN\text{-}Si) \rightarrow H\bullet(Si/contact) \quad (914.5.12)$$

(13) In both cases, the proton H^+ and hydrogen atom H• can be captured by a protonic trap denoted by X. Group-III acceptors in Si (B, Al, Ga and In discussed in section 93n), and Si• and O• in the SiO_2, at the gate/SiO_2 interface and in the Si-bulk are known protonic traps.

$$H^+ + X\bullet \rightarrow (H\bullet X)^+ \quad (914.5.15)$$

$$H\bullet + X\bullet \rightarrow H{:}X \quad (914.5.16)$$

915 Impurity Interface Trap

Electronic traps at the SiO_2/Si interface can arise from impurities introduced during processing to control the transistor characteristics. Small amounts (~1%) of chlorine, fluorine and N_2O are added to oxygen during oxidation and postoxidation anneal. Nitridation reduces oxide defects or pin-hole density, and Cl and F increase the resistance to hot electron (and hole) degradation from the larger bond strength of Si-Cl, SiO-Cl, Si-F and SiO-F than Si-H and SiO-H. Electronic interface traps should also be expected from the dopant and implanted impurities, B, In, P, and As used in current transistor designs for ULSI circuits. However, we observed only a B-related donor-like interface trap with a broad and stress-insensitive DOS located at $E_{C\text{-}Si} - 0.3$eV in boron-diffused poly-Si gate MOSC and no DOS peaks if the poly-Si gate is phosphorus-diffused. Nevertheless, the interface trap energy levels of the impurities can be estimated from the electron affinity and ionization potential of the impurity atom in vacuum, similar to Table 914.2. A list of selected impurities are given in Tables 915.1 where the values at the SiO_2/Si interface is the binding energy or energy level measured from the the Si conduction band edge. They are scaled from the vacuum value by the Coulombic dielectric (or valence-electron) screening factor $[(3.9+11.7)/2]^{-2} = 0.01644$ which overcorrects the neutral potential of the acceptor level and also the donor levels computed from the vacuum Ionization Potentials (IP) when the SiO_2/Si level becomes deeper and the orbit smaller to enclose fewer valence electrons. Thus, the estimated energy levels of the interface traps in Table 915.1 are the lower bounds.

Table 915.1
Selected List of Electron Affinity, Ionization Potential and Binding Energy of Electrons at Impurity Atoms at the Si/SiO$_2$ Interface

Vacuum					SiO_2/Si				
ELEMENT	H	N	F	Cl	Traps	H	N	F	Cl
EA	0.754	0.048	3.621	3.691	acceptor	0.012	.0008	0.060	0.061
1st IP	13.598	14.534	17.422	12.967	1st donor	0.224	0.239	0.286	0.213
2nd IP		29.601	34.970	23.81	2nd donor		0.486	0.575	0.391
3rd IP		47.488	62.707	39.61	3rd donor		0.780	1.030	0.651
4th IP		77.472	87.138	53.46	4th donor		1.273	1.432	0.878
ELEMENT	B	In	P	As	Traps	B	In	P	As
EA	0.330	0.72	0.780	0.74	acceptor	0.005	0.012	0.013	0.012
1st IP	8.298	5.786	10.486	9.81	1st donor	0.136	0.095	0.172	0.161
2nd IP	25.154	18.869	19.725	18.633	2nd donor	0.413	0.310	0.324	0.306
3rd IP	37.930	28.03	30.18	28.351	3rd donor	0.623	0.460	0.496	0.466
4th IP		54	51.37	50.13	4th donor		0.888	0.844	0.824

EA Vacuum Electron Affinity (Dickerson,Gray,Haight 1978; Huheey 1972)
OH=1.91,0.031; SiH=1.46,0.024 (Handbook Chemistry & Physics 1983)
IP Ionization Potential in vacuum (all in eV unit)

From Table 915.1, it is evident that the shallow acceptor level of all the gaseous and dopant impurities is neutral, i.e. cannot bind an electron, at room temperatures due to the small binding energy. This trend is expected of all elements in the periodic table. The first donor level from H, N, F, and Cl in Table 915.1 and O in Table 914.2 (0.224eV) are all in the $E_C - 0.2$eV to $E_C - 0.3$eV range. In this range, a donor peak associated with the oxygen dangling bond was observed which annealed at room temperature by hydrogenation [Fig.914.1(c) and (d)]. The anneal rate is expected to be different for the $E_C - 0.25$eV DOS peaks associated with N, F, and Cl as indicated by the 1st donor electron trap energy differences shown in Table 915.2. The second, third and fourth donor levels estimated by static dielectric screening for H, N, F and Cl at the SiO_2/Si interface listed in Table 915.1 are deep, so the actual values would be deeper and probably in the Si valence band and always occupied by Si valence electrons. Therefore, the 2nd, 3rd and 4th donor levels should be always electrically inactive.

Table 915.2

Selected List of Bond Energy of Impurity Atoms at the Si/SiO_2 Interface

Trap Si•	BE	Trap SiX•	BE	Trap SiO•	BE	Trap SiOX•	BE
()$_3$≡Si:H	3.101	()$_3$≡SiH	3.101	()$_3$≡SiO:H	4.430	()$_3$SiO:H	4.430
()$_3$≡Si:N	4.550	()$_3$≡SiN:H	3.519	()$_3$≡SiO:N	6.536	()$_3$SiN:H	3.519
()$_3$≡Si:F	5.728	()$_3$≡SiF:H	5.907	()$_3$≡SiO:F	2.300	()$_3$SiF:H	5.907
()$_3$≡Si:Cl	4.726	()$_3$≡SiCl:H	4.472	()$_3$≡SiO:Cl	2.819	()$_3$SiCl:H	4.472

Trap B•	BE	Trap In•	BE	Trap P•	BE	Trap As•	BE
()$_3$≡B:O	8.357	()$_3$≡In:O	3.731	()$_3$≡P:O	6.183	()$_3$≡As:O	4.99
()$_3$≡B:OH	6.394	()$_3$≡In:OH	4.050	()$_3$≡P:OH	5.306	()$_3$≡As:OH	4.71
()$_3$≡B.H.O	3.945	()$_3$≡In.H.O	3.474	()$_3$≡P.H.O	3.754	()$_3$≡As.H.O	4.039
()$_3$≡B:H	3.460	()$_3$≡In:H	2.518	()$_3$≡P:H	3.078	()$_3$≡As:H	3.648
()$_3$≡B:N	4.032	()$_3$≡In:N		()$_3$≡P:N	6.396	()$_3$≡As:N	6.032
()$_3$≡B:F	7.940	()$_3$≡In:F	1.254	()$_3$≡P:F	4.550	()$_3$≡As:F	4.250
()$_3$≡B:Cl	5.555	()$_3$≡In:Cl	4.550	()$_3$≡P:Cl	2.995	()$_3$≡As:Cl	4.64

()$_3$ (Si)$_3$ on Si side or (SiO)$_3$ on SiO_2 side of the interface.
BE Bond Energy from Handbook of Chemistry & Physics 1983-4.
Hydroxyl terminator bond energy = $(BE_{XO}+BE_{OH})/2$, BE_{OH}=4.430eV.
Hydrogen bridging bond energy = $(BE_{XH}+BE_{HO})/2$.

A similar analysis can be made on the interface energy levels from the practically important four impurities B, In, P, and As, which are listed in Table 915.1. Their electron binding energies are all smaller than those of the gaseous impurities just discussed. Thus, the second donor levels may be in the lower half of the Si gap and electrically active, although their existence have not been reported in experiments. The absence could be associated with the ease of formation of these impurity dangling bonds which can be gauged by the bonding energies given in Table 915.2 showing larger bond energy with an oxygen terminator on all four impurities than a hydrogen terminator. Thus, during fabrication, hydrogenated and unhydrogenated impurity dangling bonds are oxidized or bonded by four oxygens in the Si-substitution-site, $(SiO)_2$=X=$(SiO)_2$, which would not be easily ruptured during subsequent electrical stress.

916 Interface Trap Generation by Hot Electrons and Holes

Interface traps are created during transistor operation which determines the time to failure of both MOS and bipolar transistors. The principal source of the traps is the dehydrogenation of the electrically inactive hydrogen-terminated silicon and oxygen bonds by hot electron impact and hole capture, given by (913.3A), (913.3B), (913.4A), (913.4B), (913.6A) and (913.6B). This section gives several illustrative application examples on the new theory underlying transistor degradation due to interface trap generation by hot electrons.

Figure 916.1(a) shows the cross-sectional view of two oxide covered and gated n+/p junctions such as the source and drain junctions of a MOST or the emitter and collector junctions of a lateral BJT. The energy band diagram is given in Fig.916.1(b). These two figures show the spatial and energy tracks of the three dominant pathways of interface trap generation by hot electrons and holes. Electrons are injected from the forward-biased left-side n+/p emitter/base (BJT) or source/body (MOST) junction into the electron surface channel which is induced by a positive gate voltage. The injected electrons pass through the channel and enter the low-field or zero-field side of the space-charge region (SCR) of the reverse-biased p/n+ junction on the right-side. In this SCR, the electron is accelerated and scattered by phonons, impurities, and other electrons and holes which are represented by a combined scattering mean-free-path, λ_{elec}. It also breaks a hydrogenated interface bond when its kinetic energy (KE_{elec}) exceeds the bond energy, $qV_{BB\text{-}elec}$ or when it travels a distance $d_{BB\text{-}elect}$ (to be called the bond-breaking distance) where $KE_{elec} = qV_{BB\text{-}elect}$. The d_{BB} depends sensitively on the applied voltage through the potential variation with position in the reverse-biased SCR, which provides the major experimental test of the theory. The accelerated electron can also generate an electron-hole pair by impact when the electron kinetic energy, KE_{elec}, is greater than the pair-generation threshold energy ($E^{n}_{pn} > \approx E_{G\text{-}Si} = 1.17eV$). The impact-generated holes, labeled EIH for Electron-Impact generated Hole, are back-injected into the SCR at its maximum-field edge. These holes can be captured by the hydrogenated Si and O dangling bonds. When the hole kinetic energy KE_{hole} exceeds the hydrogen bond energy, the hydrogen is released from the Si:H or SiO:H, and the dangling-bond interface traps are activated according to (913.4A) and (913.4B). Since the space-charge layer is depleted of electrons and holes, i.e. has very low electron and hole concentrations, the released hydrogen is positively charged, or a proton H^{+}, which drifts (also diffuses) towards the source/body or emitter/base junction on the left.

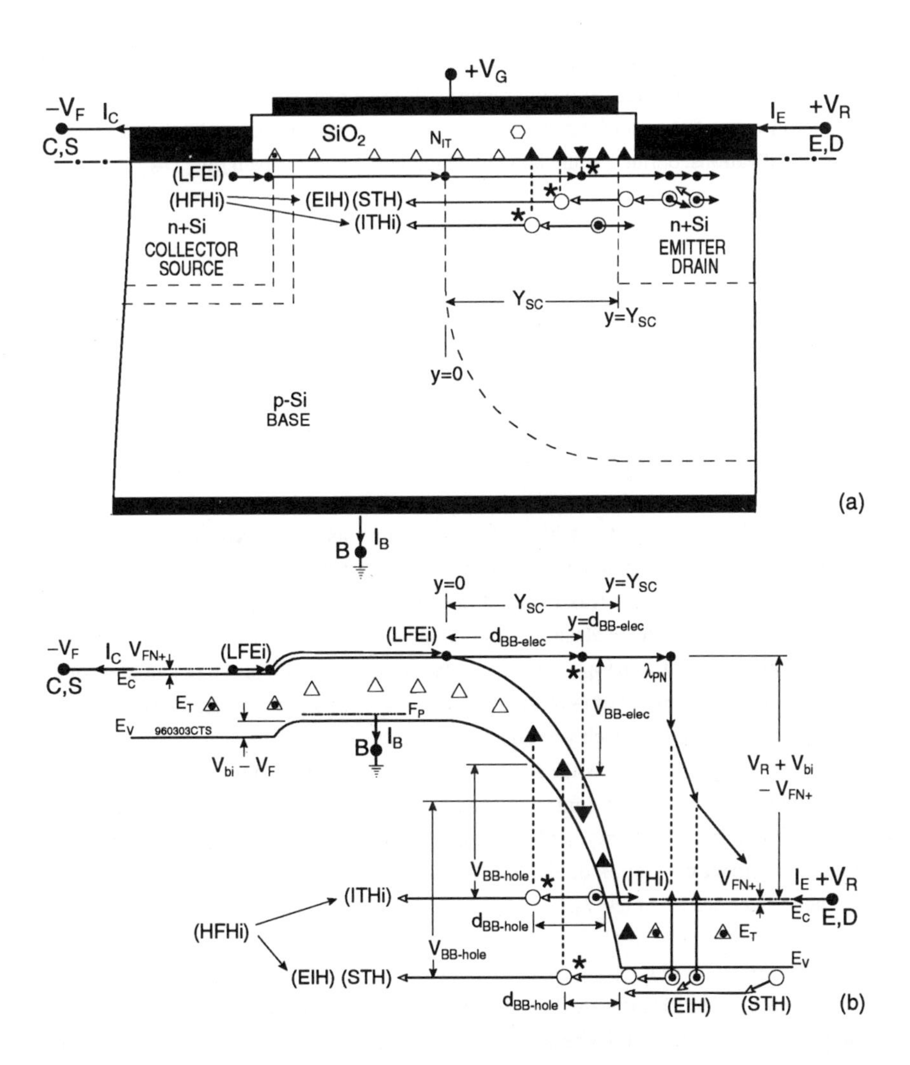

Fig.916.1 Cross-sectional view of a reverse biased p/n+ junction covered by SiO_2 showing three dominant bond-breaking pathways to generate interface traps. **(a)** Cross-sectional view. **(b)** Transition energy band diagram along the oxide/silicon interface. Bond breaking is marked by star * and vertical dash line. Open triangles are interface traps. Filled triangles are Si:H and O:H. Dots are electrons and circles are holes. Electron and hole motions are marked by arrows with solid and open tips respectively.

During the acceleration, the holes are also scattered by phonons, impurities, and other electrons and holes, to be characterized by a mean-free-path, λ_{hole}. Thus, the number of bonds broken or interface traps created per injected electron by the accelerated energetic electrons, defined as the bond breaking efficiency, is given by

$$\eta_{IT} = (dN_{IT}/dt)/(J_I/q) \quad (916.1)$$

$$\simeq q\Delta N_{IT}/J_I t \quad \text{(linear range)} \quad (916.1A)$$

$$= \int_{d_{BB}}^{Y_{SC}} \exp(-y/\lambda_3)\times(dy/\lambda_{BB}) \quad (916.2)$$

$$= (\lambda_3/\lambda_{BB})\exp(-d_{BB}/\lambda_2)\{1 - \exp[-(Y_{BB}-d_{BB})/\lambda_3]\} \quad (916.3)$$

$$\simeq (\lambda_3/\lambda_{BB})\exp(-d_{BB}/\lambda_2)\{1 - 0 \quad \} \quad (916.4)$$

$$\equiv (\lambda_3/\lambda_{BB})\cdot\eta_{IT0} \quad (916.5)$$

where $\eta_{IT0} \equiv \exp(-d_{BB}/\lambda_2)$ which is abbreviated as $\eta_{IT} = \exp(-d_{BB}/\lambda)$ in some applications. N_{IT} is the areal interface trap density. J_I is the current density or J_I/q is the electron (or hole) particle flux injected into the SCR from y=0 or $y=Y_{SC}$. The bond breaking distance which the electron (or hole) must travel to increase its kinetic energy to the bond breaking value, qV_{BB}, depends on the spatial variation of the potential through the SCR. In the integrand of (916.2), $\exp(-y/\lambda_3)$ is the fraction of injected electrons (or holes) which survived to reach y or escaped scattering and breaking a bond when moving from y=0 to y=y. (dy/λ_{BB}) is the fraction that will break a bond in the region dy. The combined mean-free-path contains three components: the surface scatterings, λ_S, containing phonon, impurity, and electron-electron and electron-hole scattering; the interband impact generation of electron-hole pairs, λ_{PN}; and the bond-breaking by the energetic electrons (or holes) λ_{BB}.

$$1/\lambda_3 = 1/\lambda_S + 1/\lambda_{PN} + 1/\lambda_{BB} \quad (916.6)$$

$$\equiv 1/\lambda_2 + 1/\lambda_{BB} \quad (916.7)$$

The length Y_{BB} in (916.3) is the larger of the hot carrier transport length (~>1000A) and the space-charge region thickness, Y_{SC}. Thus, Y_{BB} > (~1000A) >> λ_3 < (~100A) and $\exp[-(Y_{BB}-d_{BB})/\lambda_3] \ll 1$ so it can be dropped in (916.3) to give (916.4) which is just the Shockley lucky electron (or hole) model which has been used extensively for this problem, however, the most important parameter: the bond-breaking energy (or distance), was never recognized by subsequent investigators. Generally, the bond breaking mean-free-path, λ_{BB}, is large, reflecting the infrequency of the event, while λ_S is small indicating frequent scattering of the hot electron or hole.

The trap creation efficiency, (916.4), is a very strong exponential function of the applied stress voltage if the electron (hole) enters the high-field edge of the SCR. There is no voltage dependence if it enters the zero-field edge of the parabolic potential variation of the abrupt-constant p/n+ junction assumed in Fig.916.2. For the parabolic potential, the bond-breaking efficiencies of the three pathways are given as follows. Formulas for general potential and impurity variations are given in P916.1 and P916.2.

Low-Field (Zero-Field) Electron Injection (LFEi) from Base-Body Edge

$$\eta_{IT0} = \exp\{-[(2\varepsilon/qN_{AA})^{1/2}/\lambda_2]\cdot[(V_{BB})^{1/2}]\} \qquad (916.8)$$

$$d_{bb} = (2\varepsilon/qN_{AA})^{1/2}\cdot[(V_{BB})^{1/2}] = \text{independent of } V_R \qquad (916.8A)$$

High-Field (Maximum-Field) Hole Injection (HFHi) from Emitter-Drain Edge

$$\eta_{IT0} = \exp\{[-(2\varepsilon/qN_{AA})^{1/2}/\lambda_2]\cdot[(V_R+V_{bi})^{1/2} - (V_R+V_{bi} - V_{BB})^{1/2}]\} \qquad (916.9)$$

$$d_{bb} = (2\varepsilon/qN_{AA})^{1/2}\cdot[(V_R+V_{bi})^{1/2} - (V_R+V_{bi} - V_{BB})^{1/2}] \qquad (916.9A)$$

Interband Tunneling Hole Injection (ITHi) at the Emitter-Drain Edge

$$\eta_{IT0} = \exp\{[-(2\varepsilon/qN_{AA})^{1/2}/\lambda_2]\cdot [(V_R+V_{bi}-V_{Gap}-V_{FN+})^{1/2} - (V_R+V_{bi}-V_{Gap}-V_{FN+}-V_{BB})^{1/2}]\} \qquad (916.10)$$

$$d_{bb} = (2\varepsilon/qN_{AA})^{1/2}\cdot[(V_R+V_{bi}-V_{Gap}-V_{FN+})^{1/2}-(V_R+V_{bi}-V_{Gap}-V_{FN+}-V_{BB})^{1/2}] \qquad (916.10A)$$

ε is the dielectric constant of silicon. q is the electron charge. V_R and V_{bi} are the reverse applied voltage and built-in potential of the p/n+ drain/body or emitter/base (or collector/base) junction. V_{Gap} is the silicon energy gap potential. V_{FN+} is the Fermi level potential above the conduction band edge of the quasi-neutral n+drain, n+emitter or n+collector and it is zero if below. These three analytical solutions of η_{IT0} are plotted in Fig.916.2 which show four characteristics readily tested by experiments. **(i)** Electrons injected from the zero field edge, (LFEi) of (916.8). **(ii)** Holes injected from the high or maximum field edge, (HFHi) or EIH-STH of (916.9) and ITHi of (916.10) give very strong voltage dependences. **(iii)** The bond-breaking efficiency of electron injection, LFEi, is the lowest, while the higher-efficiency tunneling and thermal hole injections, ITHi and EIH-STH, differ by a voltage shift of $V_{Gap} + V_{FN+}$. **(iv)** The discrete threshold of each pathway would give a sharp structure in the η_{IT} vs V_R data which could be softened by three factors of decreasing importance: (a) V_{bi} variation due to N_{AA} and N_{DD} variation along the p/n+ junction boundary, (b) energy distribution of the impact-generated holes in EIH, and (c) thermal distribution in all pathways.

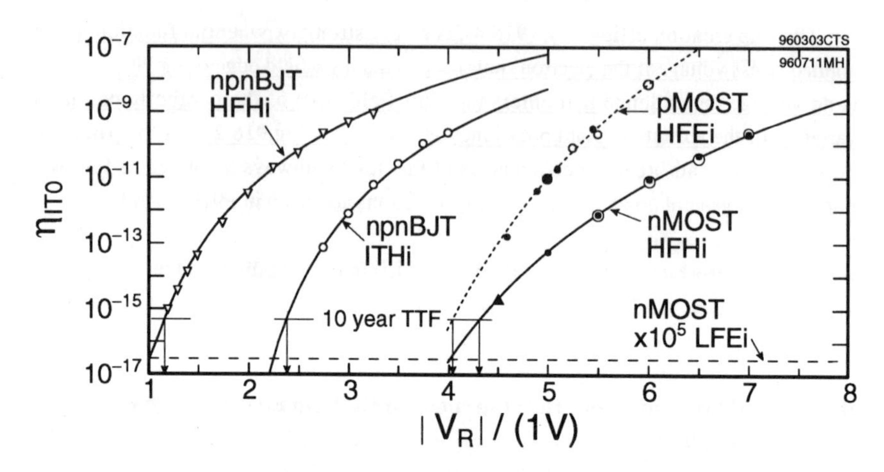

Fig.916.2 Comparison of experimental and theoretical bond-breaking efficiency by hot electrons and holes in Si npnBJT, nMOST and pMOST versus applied junction voltage. 10-year TTF is the TTF at the applied junction voltage indicated by the vertical arrow extrapolated by the theory from the TTF data at the higher V_R for 100% or 20% ΔI_B.

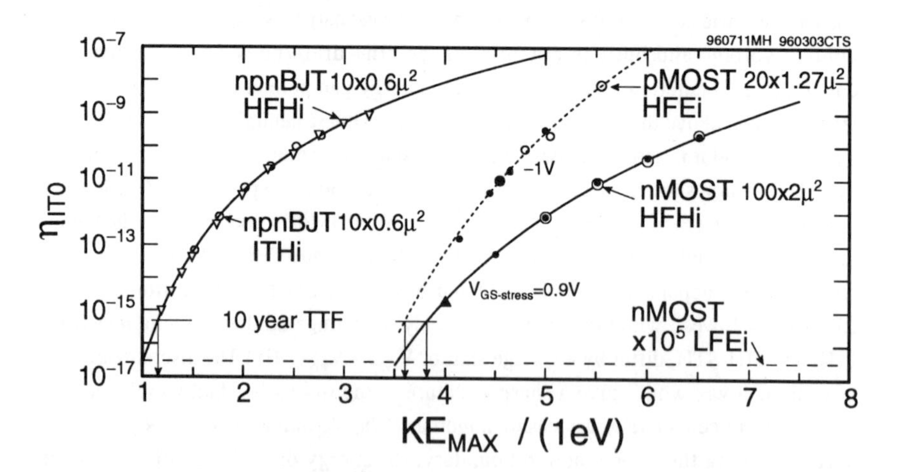

Fig.916.3 Comparison of experimental and theoretical bond-breaking efficiency by hot electrons and holes in Si npnBJT, nMOST and pMOST versus maximum kinetic energy.

This simple two-parameter theory is applied to five stress and measurement (SAM) experiments on silicon nMOST, pMOST and npnBJT to give the TTF vs V_{STRESS} which is the inverse of the bond breaking efficiency. The criteria used for the MOSTs was a 100% increase of the body current or base current, ΔI_B, in a 2-micron-technology nMOST at the peak of the I_B-V_G plot, and 20% increase in a 0.8-micron-technology pMOST. For the npnBJT, common emitter d.c. beta change by 10% was used. In the npnBJT, the stress mode was a reverse emitter/base junction voltage near breakdown with the collector/base junction open-circuited (OC), short-circuited (SC), or forward-biased (FC) to accelerate the stress by greatly increasing the stress current. The MOSTs were stressed by channel hot electrons or holes with $|V_D|>>|V_G|$ so that most of the applied drain voltage is accelerating the electrons or holes in order to give the largest build-up rate of interface traps. Stress current acceleration was attained by forward-biasing either the substrate-body-well bottom emitter junction (not shown in Fig.916.1) or the source/body junction. The efficiency data are shown in two formats. In Fig.916.2, the independent variable is the applied stress voltage across the SCR of the reverse-biased junction. In Fig.916.3, it is the maximum kinetic energy of the bond-breaking electrons or holes. The bond-breaking efficiency is not the absolute value because the thickness of the surface channel, X_{CH}, was not computed or measured so the injection current density in the definition of the bond-breaking efficiency, (916.1) J_I, was taken as the drain or emitter current divided by the drain or emitter contact area which is listed in Fig.916.3 next to each transistor. For the BJT data in Fig.916.2, the two pathways, LFEi to EIH and ITHi, differ by not only the maximum kinetic energy attainable by the holes, but also by the very large bulk-component of the emitter current from the forward-biased collector/base junction in addition to the surface channel thickness difference. These two factors are experimentally obtained by an upwards shift of the ITHi data to make its LSF curve parallel to that of the HFHi data shown in Fig.916.2. For the pMOST and nMOST, the two differences from $(\Delta I_B/I_B)_{TTF}$=100%/20%=5 and drain-contact-area=$100\text{x}20\mu m^2/20\text{x}1.27\mu m^2$=78.74 were normalized by scaling the pMOST data to the nMOST TTF condition and drain area dimension, so the remaining difference is the channel thickness which depends on the stress biases (both V_G and V_D) and the impurity concentration profile in the channel. Boron in nMOST would outdiffuse and deplete at the SiO_2/Si interface during oxidation and phosphorus or arsenic would pile up, but both profiles are modified by the ion-implant for threshold adjust and punchthrough control. Regardless of these differences, the least-squares-fits of the two-parameter theory (curves in Fig.916.2) to the data were excellent which are given below. (Many significant figures are listed for the LSF parameters, however, the data points have larger deviations.)

Table 916.3
Interface Creation via Bond Breaking by Hot Electrons or Holes
Least-Squares-Fit and Computed Parameters

LSF Parameter	nMOST	pMOST	nBJT-HFHI	nBJT-ITHI
*********************	*********	*********	*********	*********
V_{BI} (Volt)	0.8388	0.8258	1.1108	
$V_{BI}-V_{GAP}-V_{GN+}$ (Volt)				-0.1346
V_{BB} (Volt)	3.0616	3.0558	1.8767	1.8773
$\lambda_2^{-1}\sqrt{2\varepsilon_S/qN_{II}}$ (Volt)$^{-1/2}$	50.4384	71.7384	40.2342	40.2097
$V_{DS\text{-}sat}$ (Volt)	+0.50	-0.45		

Assume/Measure/Compute	nMOST	pMOST	nBJT-HFHI	nBJT-ITHI
*********************	*********	*********	*********	*********
n_i (a) (cm^{-2})	1.00×10^{10}			
kT (a) (eV)	0.02585			
λ_2 (a) (Å) holes	50		10	10
λ_2 (a) (Å) electrons		80		
N_{AA} (c) (cm^{-3})	2.03×10^{16}	1.90×10^{18}	7.95×10^{17}	7.96×10^{17}
N_{DD} (c) (cm^{-3})	6.10×10^{17}	3.90×10^{15}	5.78×10^{20}	5.77×10^{20}
N_{II} (c) (cm^{-3})	2.03×10^{16}	3.90×10^{15}	7.95×10^{17}	7.96×10^{17}

The above calculation uses the published optical-intervalley phonon scattering mean-free-paths of hot electrons (80Å) and holes (50Å) deduced from experimental interband impact generation rates of electron-hole pairs (FSSE-SG p.104) in p/n junctions in the bulk of Si. Their value near the SiO_2/Si interface could be different due to a large interfacial roughness and the different interfacial and bulk phonon modes. The calculated N_{AA} and N_{DD} in Table 916.3 depend strongly on the assumed mean-free-path value, the LSF built-in potential, V_{bi}, and the potential profile along and at the interface (assumed parabolic). If the potential profile is known, then the mean-free-path can be computed from the data rather than assumed, which would provide a measure of the interface engineering technology. Another unique LSF parameter is the bond-breaking energy. They are identical (1.88eV) along the two pathways in the npnBJT as expected. They are also identical (3.06eV) in the 2μm-nMOST and the 0.8μm pMOST technologies, although they were designed and fabricated by two different companies five years apart. This suggests that the MOST gate oxidation technology was mature and the interface had probably reached the ideal. The BJT/MOST bond-breaking energy difference could be from two factors: **(i)** the different surface impurity concentration profiles of the n+emitter/p-base, and the n+drain/p-well and p+drain/n-well from drain engineering, and **(ii)** the poorer oxide quality over the BJT-base than MOST-channel.

917 Interface Trap Creation-Destruction at High Current Density

Interface traps (Si• and SiO•) are created at the n+poly/n+emitter contact of npnBJTs by very high forward-biased electron current densities via Auger recombination given by (913.7A) and (913.7B). Their rates are proportional to N^2P which is proportional to J_E ($N=N_{DD+}$ and $P \propto J_E$) at low injection level and J_E^3 ($N=P \propto J_E$) at very high injection levels. Figure 917.1(a) shows the two trap creation pathways via breaking the hydrogen bond of Si:H and SiO:H. The SiO• comes from residual oxygen.

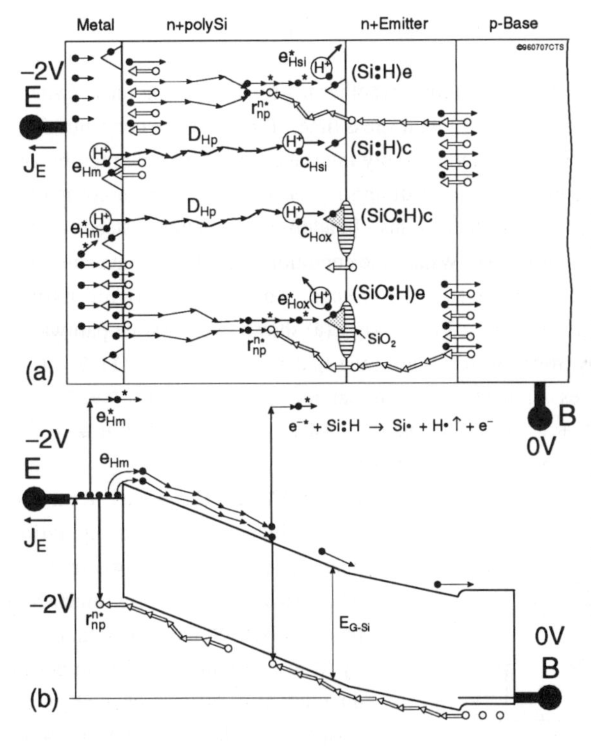

Fig.917.1 Creation-destruction of interface traps in the poly-Si n+emitter/p-base contact layers at very high forward-biased current densities. **(a)** Cross-section area. **(b)** Transition energy band diagram. [Figure (a) was first sketched by Arnost Neugroschel with inputs from C-T Sah.]

Variation of the BJT current gain, β, at very high forward emitter current densities, J_E, is due to holes, injected from the p-base into the n+emitter, recombining with the electrons at the Si• and SiO• traps in the n+poly/n+emitter interfacial layer. It follows the ideal Shockley diode law, $\exp(qV_{EB}/kT)$ at low J_E and $\exp(qV_{EB}/2kT)$ at high J_E or injection levels when $N=P>>N_{DD+emitter}$. Thus, the base current, I_B, can decrease and β can increase when hydrogen trapped at the metal/n+poly interface is released by the energetic electron generated by Auger recombination. The released hydrogen then diffuses through the n+poly layer and reaches the n+poly/n+emitter interface where it hydrogenates Si• and SiO•. The rate equation and its solution for n_{IT} at the n+poly/n+emitter interface are

$$dn_{IT}/dt = e_H(N_{ITT} - n_{IT}) - c_H H n_{IT} \quad (917.1)$$
$$n_{IT}(t) = N_{IT0}\cdot\exp(-t/\tau) + N_{IT\infty}\cdot[1 - \exp(-t/\tau)] \quad (917.2)$$
$$N_{IT\infty} = [e_H/(e_H + C_H H)]N_{ITT} \quad (917.3)$$
$$\tau^{-1} = c_H H + e_H \quad (917.4)$$
$$= c_H H + e_{H1}N_{DD+}^2 P = c_H H + e_{H1}J_E \text{ (Low } J_E.) \quad (917.5)$$
$$= c_H H + e_{H3}N^2 P = c_H H + e_{H3}J_E^3 \text{ (High } J_E.) \quad (917.6)$$

The initial value of $n_{IT}(t=0)=N_{IT0}$ determines whether n_{IT} (or I_B) will increase (when $N_{IT0}<N_{IT\infty}$) or decrease (when $N_{IT0}>N_{IT\infty}$) with stress time. The hydrogen concentration is determined by the reactions at the metal/n+poly interface which are governed by an equation similar to (917.1) but $dn_T/dt=dH/dt$ and $n_T=H<<N_{TT}$, i.e. there are many hydrogen trapping sites and few hydrogen atoms. In addition, it reaches a steady-state quickly after the current J_E is turned on. With these conditions, the steady-state solution at the metal/n+poly interface is $e_H N_{TT}/N_T=e_H+c_H H$ which is simplified using $N_T=H<<N_{TT}$ to give $H=(e_H N_{TT}/c_H)^{1/2}$. Figure 917.1(a) shows two possible pathways for release of the trapped hydrogen which control the J_E dependence of e_H. **(1)** $Si:H + 2h^+ \rightarrow Si^+ + H^+$ which gives $e_H \propto P^2 \propto J_E^2$ so that $H\propto J_E$. **(2)** $Si:H + h^+ + 2e^- \rightarrow Si:H + e^{-*} \rightarrow Si\bullet + H\bullet\uparrow + e^-$ which gives $e_H \propto N_{Metal}^2 P \propto J_E$ so that $H \propto \sqrt{J_E}$. These determine the steady-state density given by (917.3).

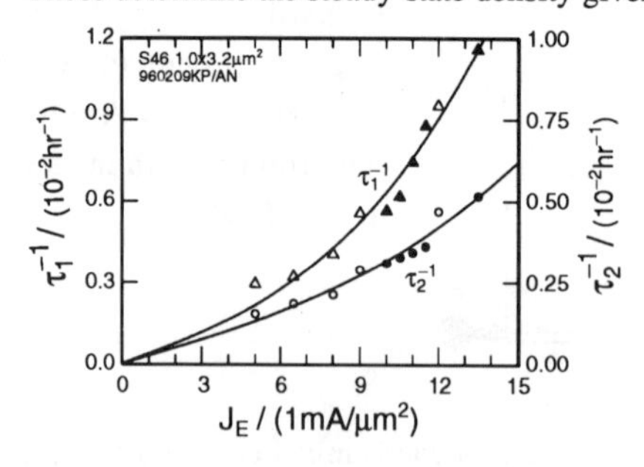

Fig.917.2 Current density dependencies of the hydrogenation-passivation rate of the two interface traps, Si• and SiO•, at the interface of the n+poly/n+Si emitter. [Data taken by M. Carroll and K. Pfaff under the direction of A. Neugroschel which was least-squares-fitted to C-T Sah's Auger hot electron generation theory.]

The LSF's are:

$1/\tau_1 = (3.62\pm0.54)10^{-4}J_E + (2.73\pm0.42)10^{-6}J_E^3$ and

$1/\tau_2 = (2.39\pm0.25)10^{-5}J_E + (7.75\pm1.99)10^{-8}J_E^3$

920 OXIDE TRAPS – Introduction

Oxide trap is an abbreviated name for the electronic traps in the SiO_2 film covering the crystalline Si of a transistor or an integrated circuit. Oxide traps originate from physical defects in SiO_2 (such as the Silicon and Oxygen Vacancies) and from impurities in SiO_2. They have electron and hole bound states with energy levels in the energy gap of the SiO_2. Those with electron bound states can trap electrons from the oxide conduction band and are called oxide electron traps. When an electron is trapped by an oxide trap in the neutral charge state, it is known as a neutral oxide electron trap. After the electron is trapped, the oxide trap becomes negatively charged. When an electron is trapped by a positively charged oxide trap, it is known as a positive oxide electron trap, and after the electron is trapped, it is neutral. Similarly, those with hole bound states are known as negative or neutral oxide hole traps. Bound states from centers with repulsive potential have not been identified in SiO_2 such as negatively charged electron trap to give the double negative charge state.

When the oxide traps are charged, they will change the electron and hole concentrations near the Si surface covered by the oxide. They will also change the electric field normal to the SiO_2/Si interface which is usually described as changing the the amount of bending of the silicon energy band near the Si surface covered by the oxide. This change of the surface electron and hole concentrations will change the recombination and generation rates at the interface traps at the SiO_2/Si interface. When the oxide charge density is high, a surface conduction channel is induced electrostatically which is detrimental to the transistor characteristics especially when it is a minority carrier channel (such as an electron channel, n-type channel or n-channel, on a p-type Si substrate). The increased electron-hole generation rate and the appearance of a channel will increase the leakage current of a p/n junction adjacent to the surface. The increased electron-hole recombination rate and the channel will increase the recombination current and decrease the BJT current gain. Oxide charges will also change the threshold voltage of the MOST, and randomly scatter electrons or holes in the channel reducing their mobility which decreases the transconductance and output current of the MOST.

This section describes the atomic origin, charging and discharging kinetics, and creation-destruction mechanisms of oxide traps. Problems and some solutions are given in section 999 at the end of this appendix.

921 Atomic Configurations and Electronic Trapping at Oxide Traps

Two energy levels have been measured by a variety of experiments in pure thin (20nm to <5nm) gate oxide films of silicon MOS capacitors and transistors. These films are thermally grown on crystalline silicon substrate in pure and dry oxygen in the latest state-of-the-art production furnaces used to manufacture multi-million-transistor integrated circuits. At the end of oxidation, the oxidized silicon wafers are suitably cooled and annealed to give a low defect density SiO_2 film and a nearly perfect SiO_2/Si interface. It is then covered by a 200-500nm boron- or phosphorus-doped low-resistance polycrystalline silicon layer which is used as the gate electrode.

The measured energy levels in the SiO_2 energy gap are: $E_C - E_A = 0.4eV$ by thermal emission of trapped electrons [1,2], $E_C - E_A = 1.0eV$ by tunnel emission of trapped electrons [2], $E_C - E_D = 7eV$ by threshold kinetic energy of electron impact emission of trapped electrons [3], and $E_D - E_V = 1.44 \pm 0.20eV$ by thermally stimulated hole emission [4]. The agreement with the independently measured (optical absorption) SiO_2 energy gap, $E_{G\text{-}SiO2} = 8.5eV = E_C - E_V \equiv E_C - E_D + E_D - E_V \simeq 7eV + 1.44 = 8.44eV$, and the nearly identical density of the electron and hole traps confirmed Sah's earlier suggestion [5] that the two energy levels are from an intrinsic defect, the amphoteric bridging oxygen vacancy center, V_O, which was characterized in detail in glass and optical fiber [5].

The oxygen vacancy possesses three charge states: the −1 and 0 charge states for the acceptor or electron trap at $E_A = E_C - 1.0eV$, and the 0 and +1 charge states for the donor or hole trap at $E_D = E_V + 1.44eV$. The atomic configurations of the three charge states and the transition energy band diagrams are shown in Fig.921.1 [5] using more accurate recent results [4]. The electron density-of-states [6] is shown in Fig.921.2.

In addition to the two energy levels of the oxygen vacancy just described, thermally stimulated hole emission experiments [4] also measured a continuous distribution of the hole trapping levels near the SiO_2 valence band, $E_T - E_V = 0.15eV$ to $1.35eV$. It has a U-shaped electronic density-of-states (DOS) versus energy. The U-shape DOS is consistent with the anticipation by Sah [8] that band edge tail states should appear from the perturbation of the SiO_2 valence band states into the SiO_2 energy gap by the random distributions of Si-O bond-length and bond-angle in the amorphous SiO_2 film.

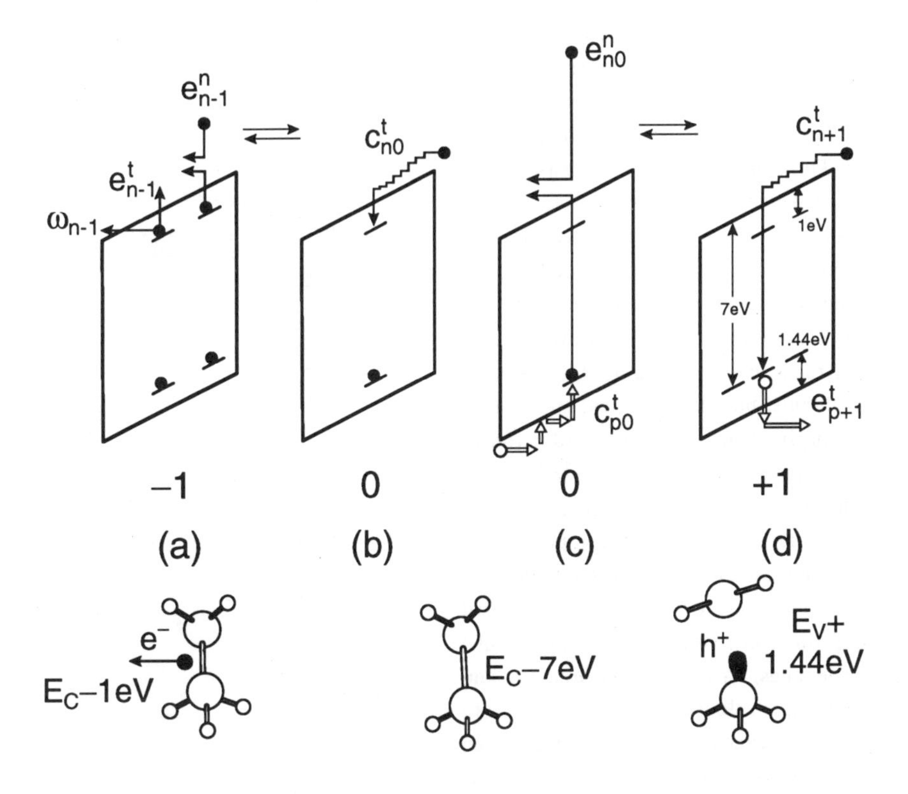

Fig.921.1 Atomic configuration and transition energy band diagrams showing the measured macroscopic electronic transition rate coefficients at the amphoteric bridging oxygen vacancy center, V_O. The black dot and e^- denote an electron and the smallest circle in the upper transition energy band diagrams and h^+ denote a hole. The larger circle is a Si^{+4} core (about 0.5A diameter as indicated by Fig.171.1 on p.94 and Slater Table 162.2 on p.92 of FSSE). The smaller circle is a O^{+6} core (1s radius is 0.068A given in Slater Table 162.2) whose four non-bonding electrons of the six $2s^2 2p^4$ valence electrons are not shown. **(a)** $V_O^{\ominus} \rightarrow V_O^{\otimes}$. **(b)** $V_O^{\otimes} \rightarrow V_O^{\ominus}$. **(c)** $V_O^{\otimes} \rightarrow V_O^{\oplus}$. **(d)** $V_O^{\oplus} \rightarrow V_O^{\otimes}$. [From Fig.B3.4 on p.422 of FSSE-SG.]

Many capture and emission transition kinetics and rate coefficients were measured for the electron trap by Thompson [3], and some for the hole trap by Thompson [3] and Yi Lu [4]. The detailed electronic density-of-states and the many, but yet-incomplete, transition rates of emission and capture of electrons and holes at the oxygen vacancy center can provide many interesting classroom-instruction and fundamental-research problems, but also practical application examples for reliability prediction in developing new manufacturing technology. This is exploited in the problems and solutions at the end of the appendix. A new interface trap generation model was just proposed by this author and experimentally verified by his students and colleagues, based on bond-breaking and hydrogen-release at the hydrogenated oxygen vacancy and non-bridging oxygen hole centers [7], shown in Fig.921.3 for the three dominant electronic traps in SiO_2. Thus, a concise microscopic picture is now available and nearly complete for the electronic traps in the VLSI and ULSI grade SiO_2 oxide film and at its interface with the underlying crystalline silicon. This complete picture, covering both the film and its interfaces, provides the model necessary for predictive and quantitative reliability designs of silicon MOS and Bipolar-Junction transistors to manufacture multi-million-transistor integrated circuit chips.

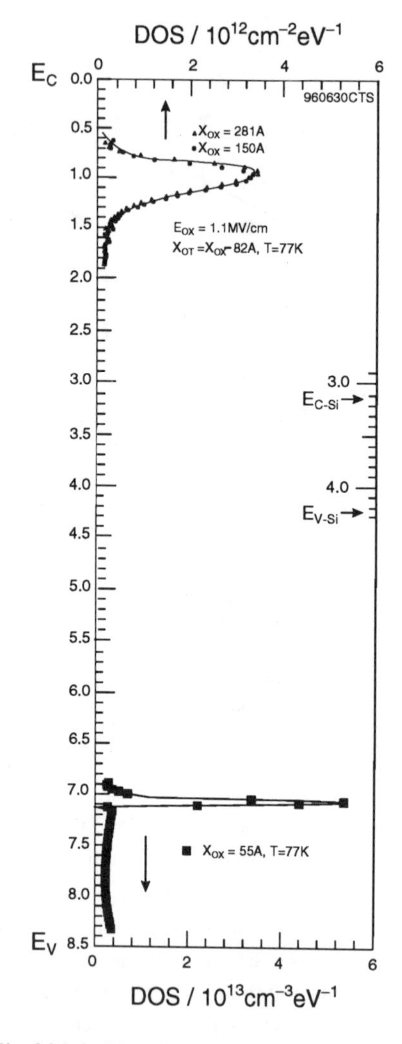

Fig.921.2 The electronic density-of-states of the two energy levels at the oxygen vacancy center, V_O, and of the random valence-band-tail states, in thin SiO_2 films [6].

The microscopic picture of the three dominant electronic traps in SiO_2 are shown in Fig.921.3. They were constructed from EPR (Electron Paramagnetic Resonance) spectra obtained on fused quartz and optical fibers, treated with a variety of temperature and ambient conditions, by Griscom and associates at the Naval Research Laboratory during the 1970's. See [5] and its cited references on these and more recent experiments. The oxygen vacancy is obtained by heating in oxygen-free ambient. The NBOHC is produced by heating the SiO_2 in a wet ambient (also called wet hole). The PR is produced by heating in oxygen at lower temperatures than the oxidation temperature to provide the excess oxygen. These generation conditions in fused quartz and optical fibers have also largely been verified with thin (5nm-30nm) gate oxide films for MOS transistors.

$(Si\text{-}O)_3{:}{:}{:}Si\bullet$	$\bullet Si{:}{:}{:}(O\text{-}Si)_3$	bridging oxygen vacancy (E' center)
$(Si\text{-}O)_3{:}{:}{:}Si\text{-}O\bullet$	$H\text{-}O\text{-}Si{:}{:}{:}(O\text{-}Si)_3$	non-bridging oxygen hole center (NBOHC)
$(Si\text{-}O)_3{:}{:}{:}Si\bullet$	$\bullet O\text{-}O\text{-}Si{:}{:}{:}(O\text{-}Si)_3$	peroxy radical (PR)

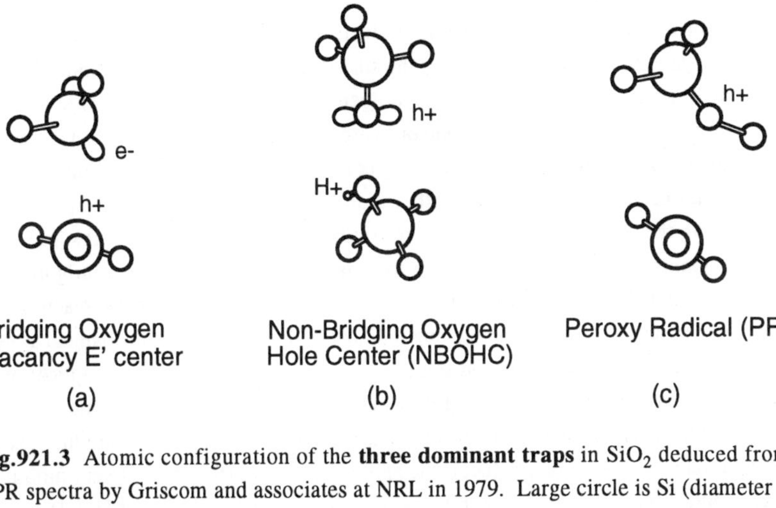

Fig.921.3 Atomic configuration of the **three dominant traps** in SiO_2 deduced from EPR spectra by Griscom and associates at NRL in 1979. Large circle is Si (diameter ~ 1A), small circle is oxygen, and very small circle with H+ is hydrogen. The bar is the two-electron bond. Ellipsoids show the shape of electron orbitals. Trapped holes are denoted by h+ and hydrogen ion or proton by H+. **(a)** Bridging Oxygen Vacancy E' center. **(b)** Non-Bridging Oxygen Hole Center (NBOHC, wet hole). **(c)** Peroxy Radical (PR). {From Fig.B3.3 on p.421 of FSSE-SG. See also reference [6].}

922 Creation and Destruction of Oxide Traps

The three dominant electronic traps in the oxide can be created and destroyed via atomic transformation by various means such as hydrogenation-dehydrogenation, oxidation, heating in vacuum or inert ambient, and exposure to ionizing radiation (low-energy x-ray and keV electrons which do not displace atoms by direct impact due to momentum conservation), in addition to nuclear radiation. We will not discuss nuclear radiation which would also create displacement damage which are additional electronic traps that would significantly complicate the simple pictures. The atomic transformation pathways and kinetics can be readily described using the chemical reaction equations. Some of these are experimentally detected or suspected to be responsible for observed experimental variation of electrical properties. They are listed below and discussed. The chemical formula in the bold bracket, [ ], is the molecular formula of the imperfection center. It contains blank spaces to indicate the vacancy from displaced or missing atoms as indicated in Figs.921.1 and 921.3. It also contains a sufficient number of adjacent atomic cores to distinguish the different electronic traps in an electrically active or electronic generic group represented by the silicon and oxygen dangling bonds, Si• and SiO•, or in an electrically inactive or molecular generic group represented by the hydrogen bond or 'silane' and OH bond or hydroxyl terminators, Si-H and Si-OH, which passivate the two electrically active dangling bonds, Si• and SiO•. This one-dimensional formula has been employed by this author to provide a simple 'one-glance' picture for visualizing the many possible atomic configurations and electronic traps and even making qualitative and semi-quantitative comparisons of the equilibrium configurations, the atomic and electronic binding and activation energies, and the atomic reaction and electronic trapping rates, without having to know or draw the three-dimensional atomic configurations. (See problems and solutions for some examples of such applications.) Room temperature is assumed in the following reactions except where indicated. Higher temperatures may speed up some reactions and retard others by increasing the reverse reaction rate. The reaction product is the atomic hydrogen, H•, which is very mobile in SiO_2 and Si even at room temperatures because of the small size of H•. It moves towards the exposed surface by diffusion as a neutral atomic hydrogen H• (or $H^{\otimes}$) and it will diffuse and also drift as a proton p^+ or hydrogen ion $H^{\oplus}$ if the electric field is high (in the p/n junction and surface space-charge layer). At the surface, two hydrogen atoms may recombine to form a hydrogen molecule, H• + H• $\rightarrow$ H_2, which then evaporates from the surface into the ambient. This movement of proton and hydrogen towards the surface is indicated by the up-arrow $\uparrow$ to indicate diffusion away from the hydrogen or proton trap to a hydrogen sink which removes the hydrogen from the electronic center in the SiO_2, lowering the hydrogen concentration continually and making the dehydrogenation reaction not reversible. However, there are also sources of high concentrations of trapped hydrogen in some parts of the Si transistor structures, such as the hydrogen in aluminum and refractory-metal-silicide contacts and interconnect layers, the Si-H in polycrystalline silicon layers, the Si•(Acceptor)•H such as $Si_3{\equiv}B{\bullet}H{\bullet}Si$ in regions highly doped by acceptor impurities (Boron-doped p+drain, p+source, and p+poly-Si gate), and in protective glasses. Even the transistor's and integrated circuit's encapsulation ambient contains hermetically sealed-in hydrogen to prevent corrosion. These sources of hydrogen could maintain a steady-state or they could reduce and even reverse dehydrogenation to stabilize the transistor and integrated circuit characteristics.

Some of the possible reactions that can destroy the oxygen vacancy, either deactivate it or convert it into one of the other two dominant centers in SiO_2, are listed in the following chemical equations. Following this equation list is another list on the creation of the Si and O dangling bonds.

Oxygen Vacancy Destruction

$[(SiO)_3{\equiv}Si{\bullet}\quad{\bullet}Si{\equiv}(OSi)_3] + 2H{\bullet} \rightarrow [(SiO)_3{\equiv}Si{:}H\quad H{:}Si{\equiv}(OSi)_3]$ (922.1A)

Oxygen Vacancy $\equiv$ Hydrogenated Si (922.1B)

$[(SiO)_3{\equiv}Si{\bullet}\quad{\bullet}Si{\equiv}(OSi)_3] + H{-}O{-}H \rightarrow [(SiO)_3{\equiv}Si{:}OH\quad H{:}Si{\equiv}(OSi)_3]$ (922.2A)

$\rightarrow [(SiO)_3{\equiv}Si{:}OH\quad{\bullet}Si{\equiv}(OSi)_3] + H{\bullet}\uparrow$ (922.2B)

$\equiv$ NBOHC (922.2C)

$[(SiO)_3{\equiv}Si{\bullet}\quad{\bullet}Si{\equiv}(OSi)_3] + 2H_2O \rightarrow [(SiO)_3{\equiv}Si{:}OH\ HO{:}Si{\equiv}(OSi)_3] + H_2\uparrow$ (922.3A)

$\equiv$ Silicon Hydroxyl (922.3B)

$[(SiO)_3{\equiv}Si{\bullet}\quad{\bullet}Si{\equiv}(OSi)_3] + \text{Hi } O_2 \rightarrow [(SiO)_3{\equiv}Si{\bullet}\quad{\bullet}O{-}O{:}Si{\equiv}(OSi)_3]$ (922.4A)

$\equiv$ Peroxy Radical (922.4B)

$[(SiO)_3{\equiv}Si{\bullet}\quad{\bullet}Si{\equiv}(OSi)_3] + \text{Lo } O_2/2 \rightarrow [(SiO)_3{\equiv}Si{-}O{-}Si{\equiv}(OSi)_3]$ (922.5A)

$\equiv$ perfect SiO_2 bond (922.5B)

Dangling Si and O Bond Creation

$$[(SiO)_3{\equiv}Si{:}H\ H{:}Si{\equiv}(OSi)_3] + 2H\bullet \rightarrow [(SiO)_3{\equiv}Si\bullet\ \bullet Si{\equiv}(OSi)_3] + 2H_2\uparrow \quad (922.6)$$
≡ Oxygen Vacancy

$$[(SiO)_3{\equiv}Si{:}H\ H{:}Si{\equiv}(OSi)_3] + 1200C \rightarrow [(SiO)_3{\equiv}Si\bullet\ \bullet Si{\equiv}(OSi)_3] + H_2\uparrow \quad (922.7)$$
≡ Oxygen Vacancy

$$[(SiO)_3{\equiv}Si{:}H\ H{:}Si{\equiv}(OSi)_3] + 2O\bullet \rightarrow [(SiO)_3{\equiv}Si{-}O\bullet\ HO{:}Si{\equiv}(OSi)_3]+H_2\uparrow/2 \quad (922.8)$$
≡ NBOHC

$$[(SiO)_3{\equiv}Si{:}OH\ HO{:}Si{\equiv}(OSi)_3]\ \text{heat} \rightarrow [(SiO)_3{\equiv}Si\bullet\ \bullet O{-}O{-}Si{\equiv}(OSi)_3]+H_2\uparrow \quad (922.9)$$
≡ Peroxy Radical

$$[(SiO)_3{\equiv}Si{-}O\bullet\ H{-}O{:}Si{\equiv}(OSi)_3]\ \text{heat} \rightarrow [(SiO)_3{\equiv}Si\bullet\ \bullet O{-}O{-}Si{\equiv}(OSi)_3]+H_2\uparrow/2 \quad (922.10)$$
≡ Peroxy Radical

$$[(SiO)_3{\equiv}SiO{:}H\ H{:}OSi{\equiv}(OSi)_3] + 2H\bullet \rightarrow [(SiO)_3{\equiv}Si\bullet\ \bullet O{-}OSi{\equiv}(OSi)_3] + 2H_2\uparrow \quad (922.11)$$
≡ Peroxy Radical

$$[(SiO)_3{\equiv}Si{:}H\ H{:}Si{\equiv}(OSi)_3] + e^{-*} \rightarrow [(SiO)_3{\equiv}Si\bullet\ \bullet Si{\equiv}(OSi)_3] + e^- + H_2\uparrow \quad (922.12)$$
≡ Oxygen Vacancy

$$[(SiO)_3{\equiv}Si{:}H\ H{:}Si{\equiv}(OSi)_3] + 2h^+ + 2e^- \rightarrow [(SiO)_3{\equiv}Si\bullet\ \bullet Si{\equiv}(OSi)_3] + H_2\uparrow \quad (922.13)$$
≡ Oxygen Vacancy

$$[(SiO)_3{\equiv}Si{:}H\ H{:}Si{\equiv}(OSi)_3] + h\nu \rightarrow [(SiO)_3{\equiv}Si{:}H\ H{:}Si{\equiv}(OSi)_3]+2e^-+2h^+ \quad (922.14A)$$
$$\rightarrow [(SiO)_3{\equiv}Si\bullet\ \ \bullet Si{\equiv}(OSi)_3] + H_2\uparrow \quad (922.14B)$$
≡ Oxygen Vacancy

The creation processes of dangling bonds or interface traps are obvious from these chemical reaction equations. For example, (922.4) and (922.5) indicate that heating the SiO_2 at high oxygen concentration will transform the oxygen vacancy into peroxy radicals while heating at low oxygen concentration will inactivate the oxygen vacancy. Reaction (922.13) indicates that the hydrogenated oxygen vacancy is activated by recombination of one of the bond electron in Si:H with an oxide hole and the recombination energy breaks the Si:H bond and releases the hydrogen which diffuses away. This oxide hole can be injected into the oxide from Si or generated by VUV light as indicated in (922.14A). **[Oxide hole** and **oxide electron,** h^+ and e^-, designate a hole and an electron in the SiO_2 valence and conduction bands respectively.**]**

923 Charging and Discharging Oxide Traps

Oxide traps can be charged and discharged by all the fundamental mechanisms occurring in the silicon bulk (section 360 and its subsections in FSSE). Because of energy conservation and the large energy gap of SiO_2 (8.5eV) compared with Si (1.12eV), and because the oxide film can be so thin in the tunneling range, some of the mechanisms are less likely than others. For example, thermal emission of trapped electrons or holes into the SiO_2 conduction or valence band is less probable from the oxide traps due to their energy depth, while the deepest trap in Si at $E_G/2 \approx 0.6$eV would give the smallest thermal emission rates of trapped electrons or holes of still 1000 electrons or holes per second at room temperature. The many tunneling transitions shown in Figs.36n0 and 36n4 of FSSE are more likely in the very thin gate SiO_2 of MOST while seldom encountered in properly designed BJT in the Si bulk. One of the major differences between SiO_2 and Si is the lack of a p/n junction in SiO_2 and the much larger energy gap of SiO_2 (8.5eV) than the visible light energy (2eV). Thus, the oxide electrons and oxide holes cannot be readily generated by conventional light sources or injected by forward biasing a SiO_2 p/n junction.

There are two other methods to inject electrons into the oxide conduction band or holes into the oxide valence band to charge the oxide traps: **(1)** by surmounting the SiO_2/Si interface potential barrier (3.12eV for electrons and 4.25eV for holes) and **(2)** by Fowler-Nordheim tunneling at high electric fields (>3.12V/X_{ox} for electrons and 4.25V/X_{ox} for holes). The tunneling pathways were described in considerable detail with many examples in section B2 of FSSE-SG, with transition energy band diagrams given in Figs.B2.1 to B2.6. The barrier surmounting methods are now discussed.

To surmount the 3.12eV-electron and 4.25eV-hole barrier at the SiO_2/Si interface, Si electrons and holes must acquire the kinetic energy. Exposing Si to deep ultraviolet light (hν > 4.25eV) is a possibility but the light is very strongly absorbed at the poly-Si gate by generating electron-hole pairs across the poly-Si energy gap (~1.12eV) and few photons are left to generate the >3.12eV electrons and >4.25eV holes at the SiO_2/Si interface. Electrical acceleration near the SiO_2/Si interface is the commonly occurring and used method to give kinetic energies larger than the SiO_2/Si barrier heights. There are four electrical acceleration methods: the a.c. and d.c. one-dimensional or normal (perpendicular) acceleration and normal injection methods known as the substrate hot electron, hole or carrier injection (SHEi, SHHi, or SHCi), and the a.c. and d.c. two-dimensional or tangential (parallel) acceleration and 90° injection methods known as the channel hot electron, hole, or carrier injection (CHEi, CHHi, or CHCi). The d.c.

methods are much easier to implement and interpret for fundamental parameter determination while the a.c. methods occur during transistor operation. Both the a.c. and d.c. 1-d SHCi methods were introduced during the early 1970's [921.9, 921.10] while the importance of the 2-d CHCi became recognized when production MOSTs' dimensions dropped below about 1-micron in 1990. These methods are summarized below.

(i) The 1-d a.c. SHCi method was popular because it employs the easily fabricated 2-terminal MOS capacitor (MOSC). It is known as avalanche electron or hole injection (AEI and AHI) in which the two-terminal MOS capacitor is pulsed into depletion so that a large accelerating potential appears through the thin surface space-charge layer. **(ii)** The 1-d d.c. SHCi method was seldom used until the mid-1980's because it requires a difficult-to-fabricate four-terminal composite transistor structure consisting of a distributively coupled MOS transistor and a vertical bipolar transistor to inject and accelerate the electrons or holes towards the SiO_2/Si interface, known as the BiMOST structure [921.9]. All production integrated-circuit MOSTs have an inherent BiMOST structure due to the p/n junction isolation well, which can serve as the emitter/base junction of the vertical BJT, and the surface space-charge layer of the MOST channel can accelerate the injected carriers. Production transistors became available to university researchers to demonstrate the BiMOST method around 1990 to assure representative, clean and nearly perfect SiO_2/Si interface attained in manufacturing. In both the a.c.-MOSC and d.c.-BiMOST structures, the surface space-charge layer must be thin or the base doping must be high (about $>5\times10^{17}cm^{-3}$) in order to accelerate a significant fraction of the injected electrons or holes to 3.15eV or 4.25eV before they are scattered by optical phonons (hot electron scattering mean-free-path ≈ 80Å, hot hole ≈ 50Å). Nevertheless, the bottom-emitter p/n junction-well in the BiMOST can be forward-biased to a sufficiently high current density to allow a significant fraction of the injected electrons or holes to escape scattering and reach the SiO_2/Si interface.

(iii-iv) The results of the 2-d d.c. CHCi methods are applicable to a.c. operating conditions because the MOST's frequency response is much faster than the charging and emission rates. In sections 91n, interface trap generation via CHC bond-breaking was discussed. During this event, 90°-scattering of the primary hot carriers (electrons in nMOST/npnBJT nBiMOST) into the oxide will charge the oxide traps. In addition, the secondary hot carriers (holes) generated by the interband impact of the primary hot carriers can also be injected into the SiO_2 by the attractive normal electric field beyond the saturation point near the drain junction where $V_D > V_G$-V_{GT}. This is an important oxide trap charging-discharging mechanism in high voltage MOSTs and BJTs when $V_D - (V_G-V_{GT})$, V_{CB} or $V_{EB\text{-reverse}}$ is larger than the SiO_2/Si barrier heights.

930 ACCEPTOR HYDROGENATION – Introduction

Hydrogenation of the group-III acceptors (A ≡ B, Al, Ga, or In) eliminates the hole bound state of the acceptor and results in an electrically inactive and neutral silicon-hydrogen-acceptor-silicon complex whose chemical formula is Si_3≡Si•H •A ≡Si_3 (or Si_3:::Si•H •A:::Si_3 in the dot notation) with hydrogen forming the bridge. This reduces the conductivity in p-type Si and increases the series resistances of the transistor. It also increases the space-charge layer thickness of the p/n junction causing changes of many d.c. and a.c. transistor parameters. The electrical deactivation of the group-III acceptors is similar to the electrical deactivation of the interface traps by hydrogen discussed in sections 91n but with an important fundamental microscopic difference: the hydrogen serves as the terminator for the Si• and SiO• dangling bonds of the interface traps while in acceptor deactivation, the hydrogen is the bridge.

Hydrogenation of the group-III acceptors was first identified by Sah, Sun and Tzou in 1982 [1]. A comprehensive analysis of the hydrogenation-dehydrogenation kinetics including hydrogen transport in the oxide layer and its two interfaces was later given by Sah, Pan and Hsu [2]. Sah introduced the concept and coined the term **proton trap** [2,3,4]. The general term is the protonic traps [2] analogous to the electronic (or electron and hole) traps, however, antiproton trap or a proton-hole trap does not exist in materials under transistor operation conditions or energies. Thus, the group-III acceptors in Si are not only an electronic trap (in this case, the hole trap with hole bound state and not the electron trap with electron bound state) but also an atomic trap for proton or hydrogen atom. Other defects (Si and O dangling bonds, and vacancies) and impurities in Si are also potential or established protonic traps. A detailed historical survey was recorded to show the dates and sequence of each historical fundamental experiment and idea that formed the basis of the **hydrogen bridge** which was a well-known concept and model in classical chemistry [3,5,6].

The most abundant hydrogen sources are the interfaces (SiO_2/Si and poly-Si/c-Si) and grain boundaries of the polycrystalline silicon film, and the amorphous silicon film sometimes used in place of the polycrystalline silicon film. If the oxide layer is grown in wet ambient or in the presence of hydrogen, it will also contain a high concentration of hydrogen. It was determined by experiments that the hydrogen can be released from its bonding site by a low current density of electrons with kinetic energy greater than the hydrogen bond energy which is <3.5eV. Four means to produce the energetic electrons were demonstrated (three by us) which release the hydrogen from the hydrogen sources in the surface layers. The released hydrogen then diffuses to the silicon substrate to electrically deactivate the boron and acceptor in the silicon substrate of MOS capacitor or transistor structures. The first two listed below are encountered during transistor operation and the last two, during transistor fabrication of future sub-0.1-micron dimension: **(1)** transient (or avalanche) and steady-state substrate hot electron injection (SHEi), **(2)** tunnel electron injection with acceleration by the oxide electric field, **(3)** keV electron exposure, and **(4)** exposure to x-ray. **(5)** In addition, hydrogen release and hydrogenation were observed at very high current densities when the emitter of a BJT is forward biased to very high current densities ($>\sim 10^6 A/cm^2$).

931 Atomic Configuration of Hydrogenated Acceptors

The atomic configuration of the hydrogen-acceptor complex in silicon (and other semiconductors) was an attraction to the fundamental research physicists because it is the simplest system containing more than one atom. The microscopic picture was deduced from infrared and bound exciton absorption spectra including also deuterium substitution to observe the mass effect. The experimental results were corroborated with theoretical cluster calculations. The final consensus model on the precise spatial location of the hydrogen and its adjacent silicon atoms was exactly the original 3-center/2-electron bond-center bridging hydrogen model proposed by us [1,3], in which the hydrogen is located near the center of the B-Si bond as indicated in the Figs.931.1(a), (b) and (c) with a chemical formula of $Si_3 \equiv Si \bullet H \bullet B \equiv Si_3$, rather than the 2-center/2-electron Pankove model of Fig.931.1(d) in which the hydrogen is attached to the Si and the boron is pulled back to form a planar sp^2 orbital $Si_3 \equiv S{:}H\ B \equiv Si_3$. Based on simple electrostatics (Coulomb repulsion between two positive charge), the concept of bond length and strength, and configuration symmetry and degeneracy (degeneracy lowers the total energy), we also considered in 1982 all the various lattice reconstructions of the 9-atom clusters (7-Si, 1-B and 1H) which was summarized in the 1984 SRC-Contract Annual Report, presented and its figures distributed at the 1985-1986 Gordon Conferences, and discussed in the 1988 IEE handbook review (See p.592 of reference [3].) Their (110) plane view are given in Figs.931.1-931.3. Figure 931.1(a) shows the restricted case of bond-center H^+ and undisplaced B^{+3}. Figure 931.1(b) is the general case with all six atoms are displaced. Figure 931.1(c) is probably the lowest energy configuration.

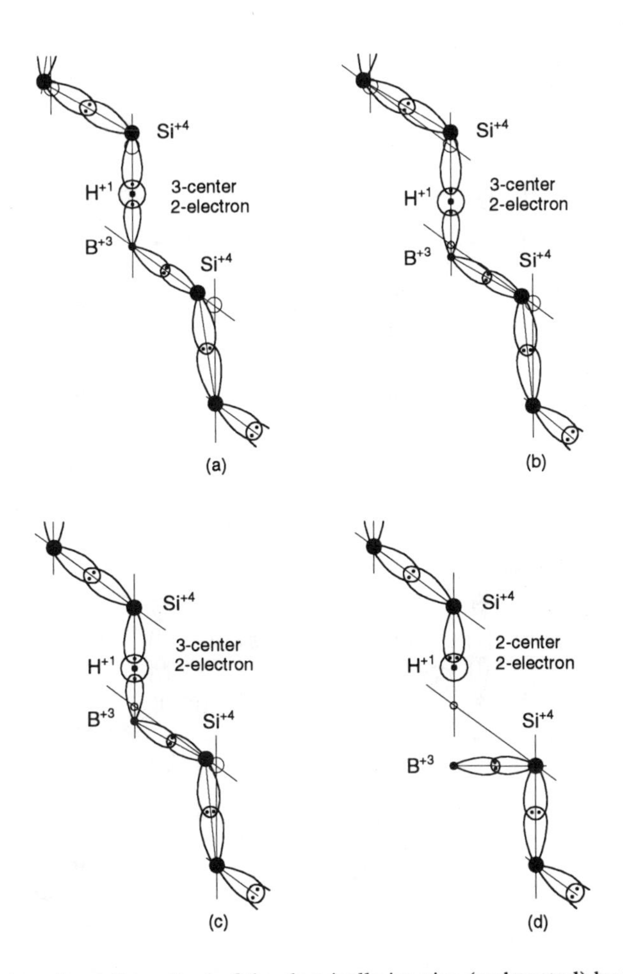

Fig.931.1 Atomic configurations of the electrically inactive (and neutral) hydrogenated boron acceptor in the (110) Si plane showing atomic displacement from the diamond lattice points. (a)-(c) have the **hydrogen bridge**, Si•H •B ≡Si_3, and are 3-center/2-electron centers. (d) has the **hydrogen terminator**, Si-H B≡Si_3, and is a 2-center/2-electron center. **(a)** B^{+3} not displaced. All four adjacent Si^{+4} displaced. **(b)** B^{+3} and adjacent Si^{+4}'s all displaced. **(c)** (B^{+3})-$(Si^{+4})_3$ displaced and coaxial. Si^{+4} not displaced. **(d)** (B^{+3})≡$(Si^{+4})_3$ detached from H^+ to form planar B≡Si_3 sp^2 orbital. Si• terminated by H•. (Drawn approximately to scale.) Adapted from C.-T. Sah cited in references [4,5,6].

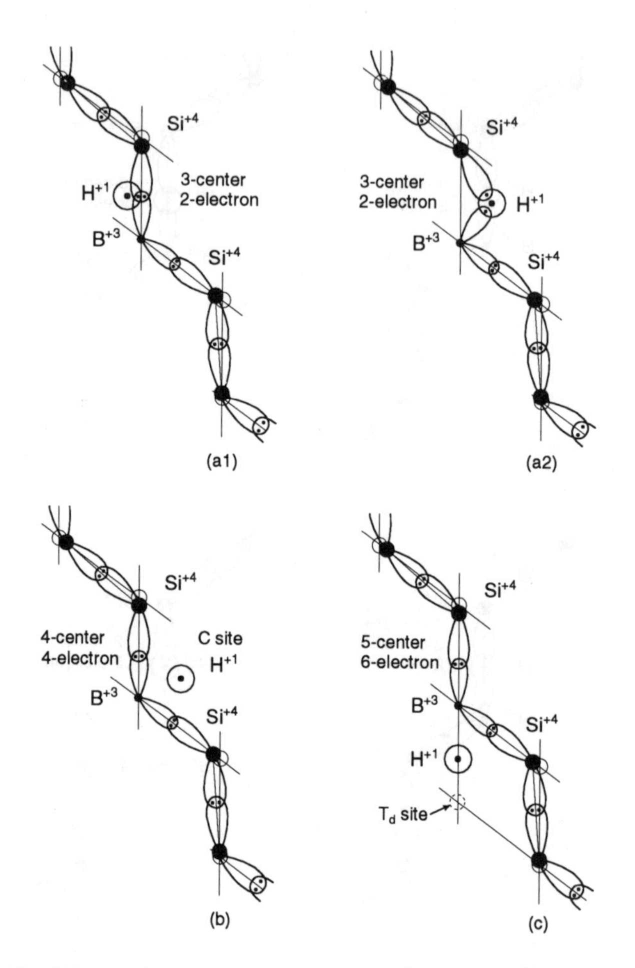

Fig.931.2 Possible atomic configurations of the hydrogenated boron acceptor in the (110) silicon plane with hydrogen outside of the Si-B bond. **(a1)** and **(a2)** 3-center/2-electron. **(b)** 4-center/4-electron with $H^{\otimes}$in the C-site. **(c)** 5-center/6-electron with $H^{\otimes}$ next to the T_d-site. (Drawn approximately to scale.) Adapted from C.-T. Sah cited in references [4,5,6].

The other atomic configurations, with H outside of the Si-B bond, were proposed and eventually all rejected by experiments or theoretical cluster calculations. These are the two other 3-center/2-electron configurations with H^+ displaced out of the Si-B bond as shown in Figs.931.2(a1) and (a2), the 4-center/4-electron configuration with the interstitial H^+ at the C-site in Fig.931.2(b), and the 5-center/6-electron configuration with the interstitial H^+ at the T_d-site. The bond-strength of the eight hydrogenated boron acceptor configurations were estimated by Sah [3] and listed in Table 931.2.

Table 931.2
Estimated Chemical Shift of the Hydrogen Bond Strength

```
931.1(a) >931.1(c) >931.1(d)  >931.2(a1) >931.2(a2)>931.2(b) >931.2(c)
931.1(b)                                             Si Si       Si
                                  H         H         \ /         |
 Si•H•B  > Si•H• B >Si-H B-Si >  Si-B    >  Si•  •B > Si-B-Si > Si-B-Si
                                                       H          H
 3.22eV  >  3.2eV  >  3.1eV   >   ~2eV   >  ~1.7eV > ~1.4eV  > ~1.1eV
```

The rejected configurations all have lower symmetry, fewer equivalent H^+ positions or lower configuration degeneracy, and weaker hydrogen binding energy than the three bridging hydrogen configurations in Figs.931.1(a), (b) and (c). The displaced B^{+3} and its three adjacent Si^{+4} of Fig.931.1(c) has probably the lowest total energy because its lattice reconstruction energy is the smallest during proton (H^+) tunneling from $Si{\bullet}(H^+){\bullet}B \equiv Si_3$ to one of the three adjacent B-Si bridging sites in $B{\equiv}Si_3$, involving only the Si^{+4} displacement of the destination B-Si bond while (a) and (b) involve the displacement of all four nearest neighbor Si^{+4} and two of the three Si^{+4} cores in the second-nearest-neighbor $(Si^{+4})_3$.

The atomic configuration of each hydrogenated acceptor of the entire group-III acceptors are shown in Fig.931.3(a)-(d). Although more complicated than boron because of the large core radius, they are of increasing technological importance. For example, IBM reported (1996) the use of indium, the slowest diffusing acceptor in Si, in the p-base-well of experimental 0.07-micron nMOST to prevent drain-to-source punch through. The chemical shift of the bond energies in vacuum are listed in Table 931.2 [3] which may be used as a guide for estimating the stability of the complex in silicon. This vacuum bond energy is reduced in Si and SiO_2 by the adjacent positively charged silicon and oxygen cores and bond electrons as well as by distant valence electron screening.

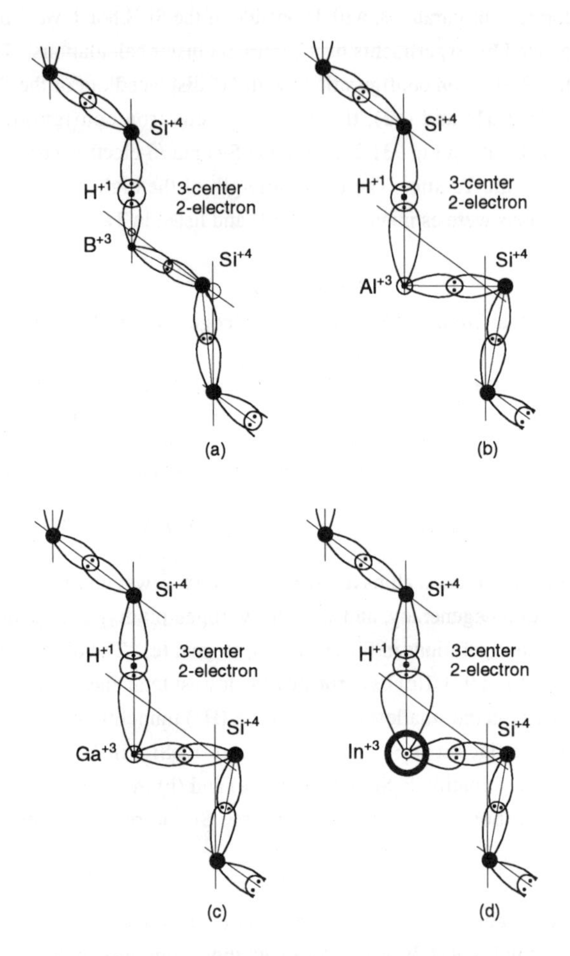

Fig.931.3 Atomic configuration of the hydrogenated group-III acceptors (B,Al,Ga,In) in the (110) silicon plane with the 3-center/2-electron hydrogen bridge configuration represented by $Si_3 \equiv Si \bullet H \bullet A \equiv Si_3$. **(a)** B, **(b)** Al, **(c)** Ga, and **(d)** In. (Drawn approximately to scale.) Adapted from C.-T. Sah cited in references [4,5,6].

Table 931.2
Bond Energy of Silicon and Hydrogen Bonds in Vacuum

IMPURITY A=	Si-A (eV)	A-H (eV)	Si•H•A (eV) Table	Data
********	******	******	******	******
O	8.39	4.43		
H	3.10	4.52		
Si	3.39	3.10	3.25	
B	2.98	3.46	3.22	1.07
Al	2.60	2.95	2.88	2.27
Ga		<2.84	2.5	2.22
In		2.52	2.2	
Tl		1.95	2.	

932 Acceptor Hydrogenation Kinetics

Some of the acceptor hydrogenation and dehydrogenation reactions are listed by the chemical reaction equations indicated below where boron acceptor is assumed while the reactions also apply to the other group-III acceptors. For an example of the detailed individual steps, one of which may dominate the observation in an experiment, see the problems and solutions.

Dehydrogenation

$$Si_3{\equiv}Si{\bullet}H{\bullet}B{\equiv}Si_3 + \text{heat} \rightarrow (Si_3{\equiv}Si\text{-}B{\equiv}Si_3)^- + h^+ + H{\bullet}\uparrow \qquad (932.1)$$
$$Si_3{\equiv}Si{\bullet}H{\bullet}B{\equiv}Si_3 + h^{+*} \rightarrow (Si_3{\equiv}Si\text{-}B{\equiv}Si_3)^- + H^+\uparrow + h^+ \qquad (932.2)$$
$$Si_3{\equiv}Si{\bullet}H{\bullet}B{\equiv}Si_3 + e^{-*} \rightarrow (Si_3{\equiv}Si\text{-}B{\equiv}Si_3)^- + H^+\uparrow \qquad (932.3)$$
$$Si_3{\equiv}Si{\bullet}H{\bullet}B{\equiv}Si_3 + 2h^+ + e^- \rightarrow Si_3{\equiv}Si{\bullet}H{\bullet}B{\equiv}Si_3 + h^{+*} \qquad (932.4A)$$
$$\rightarrow (Si_3{\equiv}Si\text{-}B{\equiv}Si_3)^- + H^+\uparrow + h^+ \qquad (932.4B)$$
$$Si_3{\equiv}Si{\bullet}H{\bullet}B{\equiv}Si_3 + h^+ + 2e^- \rightarrow Si_3{\equiv}Si{\bullet}H{\bullet}B{\equiv}Si_3 + e^{+*} \qquad (932.5A)$$
$$\rightarrow (Si_3{\equiv}Si\text{-}B{\equiv}Si_3)^- + H^+\uparrow \qquad (932.5B)$$
$$Si_3{\equiv}Si{\bullet}H{\bullet}B{\equiv}Si_3 + H{\bullet} + e^- \rightarrow (Si_3{\equiv}Si\text{-}B{\equiv}Si_3)^- + h^+ + H_2\uparrow \qquad (932.6)$$

Hydrogenation

$$(Si_3{\equiv}Si\text{-}B{\equiv}Si_3)^- + H{\bullet} + h^+ \rightarrow Si_3{\equiv}Si{\bullet}H{\bullet}B{\equiv}Si \qquad (932.7)$$

940 ELECTROMIGRATION – Introduction

As the dimension of silicon integrated circuit decreases to below one-micron, the interconnect metal lines which electrically connect the adjacent transistors become narrower and thinner while the current does not decrease proportionally so the current density increases. At high current densities, the large number of electrons in the interconnect metals lines will impart sufficient energy and momentum to cause the metal atoms (charged atomic cores) to migrate by diffusion and drift in the direction of the electrons. This would produce voids in the metal line, reduce its cross-sectional area, increase the electron current density, and increase the resistance and local heating, all of which would accelerate the migration because of higher energy and momentum transfer by the larger number of electrons colliding with the atoms. This regenerative sequence of events would eventually open up the metal line and cause electrical short-circuit to adjacent conductors separated by an SiO_2 insulator due to melting through the insulator. The cross-sectional picture of a nMOST shown in Fig.940.1 illustrates the beginning portion of the two aluminum interconnect lines, labeled L_{AL}, one to the gate (the Word Line of a SRAM or DRAM cell such as that shown in Figs.671.1-671.2 and Figs.673.1-673.3) and one to the drain (the Bit Line of a DRAM or SRAM cell).

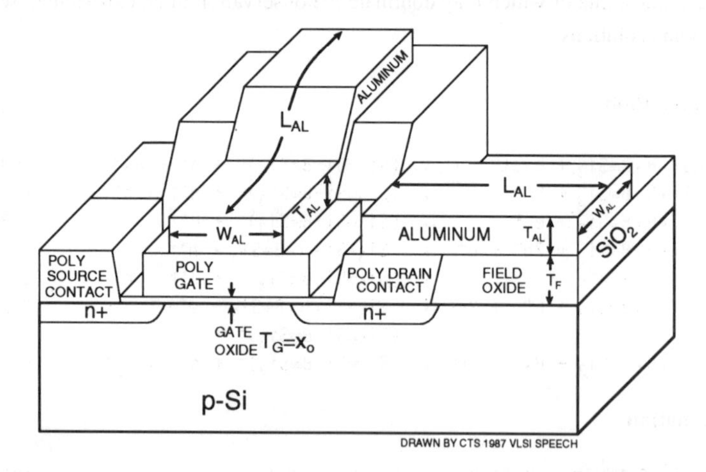

Fig.940.1 The cross-sectional view of a Si nMOST showing the two aluminum interconnect lines to the gate and the drain labeled L_{AL}. The aluminum interconnect line to the source is not shown.

941 Empirical Characterization of Electromigration

Electromigration has been characterized by an empirical equation of Mean-Time-to-Failure (MTF) known as the Black equation given by

$$MTF = MTF_0 J^{-2} \exp(E_A/k_BT) \tag{941.1}$$

J is the current density. E_A is the thermal activation energy which is 0.5eV for grain-boundary diffusion and 1.4eV for self-diffusion of the aluminum in the bulk crystalline aluminum metal. Three factors will affect the activation energy and the pre-exponential factor, MTF_0, which are: (i) the impurities in the aluminum line, such as the residual oxygen and hydrogen, and the intentionally added hardening impurity such as Cu to improve the electromigration resistance, (ii) thickness of the interconnect line and the proximity of the two SiO_2/Al boundary interfaces, and (iii) considerable variations and distributions of the grain size and grain-boundary abundance from fabrication. A macroscopic equation would give the local macroscopic resistivity that models the void growth resulting from the migration of the Al atom from a void-seed site to an Al-sink, and a finer-macroscopic equation would give the rate of increase of the void size due to the high density of electrons flowing tangentially next to the void's surface which break the metallic surface bond (electron bond) and move the aluminum atoms on the void's surface downstream with the electrons, causing the void to grow. Such a surface layer peeling bond-breaking model has not been analyzed by solving the electron diffusion and electron flux equations with a mobile aluminum generation term to characterize the aluminum-bond (metallic bond) breaking on the void surface. (See problem P941.1.)

942 Circuit Performance Limited by Interconnect Delay

Electromigration failure limit is intimately related to the switching speed limit of digital integrated circuits from the resistance-capacitance delay of the interconnect line. For a given linewidth or technology, the line's series resistance and the current density are inversely proportional to the thickness of the metal line. Thus, the switching speed and MTF are directly proportional respectively to the thickness and thickness squared. However, the thickness is limited by the aspect ratio of the gas-plasma etched channel or groove due to: thickness ≡ channel depth ≃ channel width = limited by lithography. The RC interconnect or wire delay of Fig.940.1 is given by

$$t_d = RC = (\rho_W L_W / T_W W_W) \cdot (\varepsilon_{ox} W_W L_W / T_F) = \rho_W \varepsilon_{ox} (L_W^2 / T_W T_F) \tag{942.1}$$

$$t_d / L_W^2 = 93\text{ps/mm}^2 \quad \text{(Aluminum Wire)} \tag{942.2}$$

where subscript W stands for wire, ρ_W=2.7μΩ-cm, ε_{ox}=3.9×88.54fF/cm, L_W=1.00mm, T_W=0.1μm, and T_F=0.1μm. The trend of interconnect-delay limit of logic integrated circuit speed was evident even in the first generation microprocessor chip design as indicated in Fig.942.1. It shows the characteristic L^2 dependence for the earlier generation microprocessor (>1-micron) technologies that would have implied extremely low electron mobility. Thus, the L^2 dependence must come from interconnect delay of a line connecting 128 gates rather than the true channel length delay. This is tabulated in the Table 942.1 which used t_{RC}=93(ps/mm^2)L_W^2=t_{data}, A_{MOST}=40L_G^2 (i.e. 40 squares per transistor with LDD), and N=$L_W/\sqrt{A_{MOST}}$ (number of gate or MOST per L_W).

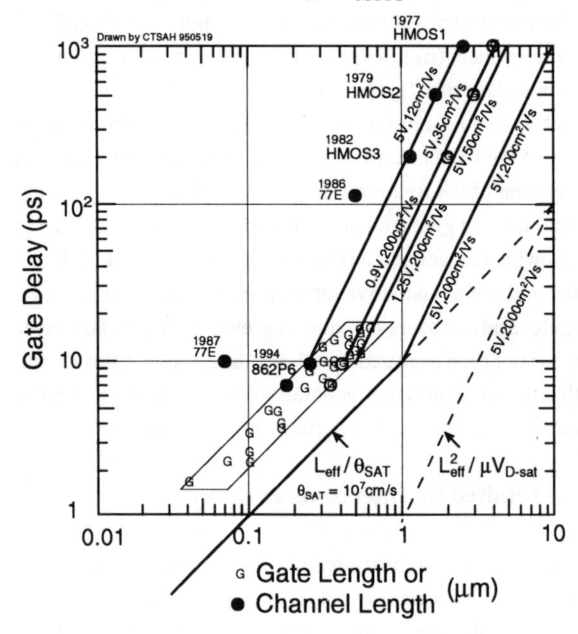

Fig.942.1 Gate delay as a function of gate length or channel length, showing characteristic L^2 dependence predicted by Equation (942.1) and the L^1 dependence due to velocity saturation. Note the unreasonably low mobilities for the early generation > one-micron technology.

Table 942.1
Interconnect Wire Delay in Intel Microprocessors

Generation	L_G (μm)	t_{data} (ps)	L_W (μm)	A_{MOST} (μm^2)	$\sqrt{A_{MOST}}$ (μm)	N MOST/L_W
1977 HMOS1	4	1000	3279	640	25.3	130
1979 HMOS2	3	500	2318	360	19.0	122
1982 HMOS3	2	200	1466	160	12.6	116
1994 862P6	0.4	9.5	320	6.4	2.53	126

999 BIBLIOGRAPHY AND REFERENCES

[911.1] Chih-Tang Sah , "Interface traps on Si surfaces," in Properties of Silicon, section 17.1 pp.499-507, INSPEC, the Institution of Electrical Engineers, London, New York, May 1988.

[911.2] Chih-Tang Sah, "Interface traps on oxidized Si from electron spin resonance," section 17.2 in Properties of Silicon, pp.508-511, INSPEC, the Institution of Electrical Engineers, London, New York, May 1988.

[911.3] Chih-Tang Sah, "Interface traps on oxidized Si from x-ray photoemission spectroscopy, MOS diode admittance, MOS transistor and photogeneration measurements," section 17.3 in Properties of Silicon, pp.512-520, INSPEC, the Institution of Electrical Engineers, London, New York, May 1988.

[911.4] Chih-Tang Sah, "Interface traps on oxidized Si from two-terminal dark capacitance-voltage measurements on MOS capacitors," section 17.4 in Properties of Silicon, pp.521-531, INSPEC, the Institution of Electrical Engineers, London, New York, May 1988.

[914.5] Chih-Tang Sah, "Studies of the Reliability Physics of Silicon VLSI Transistors," Annual Report No. 1 (01 May 1983) to 31 July 1984) of Contract No. SRC 83-01-030 and SRC Technical Report Series TRS-100, Semiconductor Research Corporation, P.O.Box 12053, Research Triangle Park, NC 27709 USA. 97pp. Chapter II of this report contains comprehensive treatment of traps on oxidized VLSI silicon including fundamentals, mechanisms amd methods of trap generation and annealing, measurement or characterization methods, and kinetics analyses of data. Oxide and interface traps, and acceptor hydrogenation are described from which the background summary in this appendix is abstracted.

[914.6] Chih-Tang Sah, Jack Y-C Sun and Joe J-T Tzou, "Generation-annealing kinetics of interface states on oxidized silicon activated by 10.2eV photohole injection," J.Appl.Phys. v53, 8886-8893, December 1982.

[921.1] Tak H. Ning, "Thermal re-emission of trapped electrons in SiO_2," J. Applied Physics, 49(12), 5997-6003, December 1978.

[921.2] Scott E. Thompson and Toshikazu Nishida, "Tunneling and thermal emission of electrons from a distribution of shallow traps in SiO_2," Appl. Phys. Lett. 58 (12), 1262-1264, 25 March 1991.

[921.3] Scott E. Thompson and Toshikazu Nishida, "Positive charge generation in SiO_2 by electron-impact emission of trapped electrons," J. Appl. Phys. 72(10), pp.4683-4695, 15 November 1992.

[921.4] Yi Lu and Chih-Tang Sah, "Thermal emission of trapped holes in thin SiO_2 films," J. Appl. Phys. 78(5), 3156-3159, 1 September 1995.

[921.5] Invited plenary talk by Chih-Tang Sah, "Mechanisms of electron trapping in SiO_2," 21st International Conference on the Physics of Semiconductors, Beijing, China, 10-14 August 1992. Vol.1, pp.28-40, World Scientific Publishing Co. Singapore. The article in the conference proceedings was written by Sah and the talk was delivered by Nishida from Sah's taped speech because of the illness of a family member which prevented Sah from traveling to Beijing. For a figure, see Fig.B3.4, p.422, in FSSE-SG.

[921.6] This electronic density-of-state picture was first presented by Sah in a talk titled "Fundamental electronic degradation pathways in MOS gate oxides," on 29 September 1994 in a Technology Seminar at the Motorola Semiconductor Sector, Phoenix, Arizona. It was subsequently presented in three seminars at three silicon integrated circuit manufacturers, IBM Research Center on 14 October 1994, AMD-Sunnyvale on 21 April 1995, and AT&T Microelectronics-Orlando on 27 June 1995.

[921.7] Atomic configuration picture of the three dominant electronic traps in SiO_2 are given in Fig.B3.3 on p.421 of FSSE-SG and also given in the Beijing invited talk cited in [5].

[921.8] Chih-Tang Sah, "Origins of interface states and oxide charges generated by ionizing radiation in thermally oxidized silicon," IEEE Transaction on Nuclear Science, NS23(6), 1563-1568, 1 December 1976.

[921.9] Charles C. H. Hsu, Toshikazu Nishida, and Chih-Tang Sah, "Observation of threshold oxide electric field for trap generation in oxide films on silicon," J. Applied Physics, 63(12), 5882-5884, 15 June 1988.

[921.10] Toshikazu Nishida, "BiMOS and SMOSC structures for MOS parameter measurements," Solid-State Electronics, 35(2), 357-369, March 1992.

[921.11] Kun Huang in Chinese. See English presentation given by Ming-Fu Li in **Modern Semiconductor Quantum Physics**, sections 380 and 391-394.

[921.12] Alex C. K. Wang, Luke Su Lu and C. T. Sah, "Electron capture at the two acceptor levels of a zinc center in silicon," Phys. Rev. B, v30(10), 5896-5903, 15 November 1984.

[921.13] Jack T. Kavalieros and Chih-Tang Sah, "Separation of interface and nonuniform oxide traps by the DC current-voltage method," IEEE Trans. Electron Devices, v41(1), 137-141, January 1996.

[930.1] Chih-Tang Sah, Jack S-C Sun and Joe T. Tzou, "Deactivation of boron acceptor in silicon by hydrogen," Appl. Phys. Letts. v43, 204-106, 15 July 1983. "Deactivation of group III acceptors in silicon during keV electron irradiation," Appl. Phys. Letts. v43(10), 962-964, 15 November 1983.

[930.2] Chih-Tang Sah, Sam S-C Pan and Charles C-H Hsu, "Hydrogenation annealing kinetics of group-III acceptors in oxidized silicon," J. Appl. Phys. v57(12), 5148-5161, 15 June 1985.

[930.3] Chih-Tang Sah, "Hydrogenation and dehydrogenation of shallow acceptors and donors in Si: fundamental phenomena and survey of literature," section 17.16 in <u>Properties of Silicon</u>, pp.584-604, INSPEC, the Institution of Electrical Engineers, London, New York, May 1988. ibid., et.al., "Hydrogenation and dehydrogenation of shallow acceptors and donors in Si: fundamental phenomena and survey of the literature,", section 17.17 on pp.604-612. For original figures and summary of literature data, see also the two SRC Annual Reports cited as references [13] for AR No.1 and [18] for AR No.2 in section 17.16 on p.600-601 of <u>Properties of Silicon</u>.

[930.4] Chih-Tang Sah, "Studies of the Reliability Physics of Silicon VLSI Transistors," Annual Report No. 2 (01 August 1984 to 31 July 1985) of Contract No. SRC 83-01-030 and SRC Technical Report Series T85076, Semiconductor Research Corporation, P.O.Box 12053, Research Triangle Park, NC 27709 USA. 22pp. This report contains a detailed account of the discovery and followup experiments of hydrogenation and dehydrogenation of group-III acceptors in silicon. Some of which are also described and updated in the two sections in <u>Properties of Silicon</u> cited in reference [930.3].

[930.5] Chih-Tang Sah and S.C. Pan, "Summary of chronology of boron and group-III acceptor hydrogenation in crystalline silicon," Gordon Research Conference on <u>Line Defects and Interfaces in Semiconductors</u>, Plymouth State College, NH, USA. July 8-12, 1985.

[930.6] Chih-Tang Sah and S.C. Pan, "Experiments and model on deactivation of boron and other group-III acceptors at the Si-SiO_2 interface," Gordon Research Conference on <u>Metal Insulator Semiconductor Systems</u>, Tilton School, Tilton, New Hampshire, USA. July 14-18, 1986.

References for section 931 are identical to those for section 930.

[931.1] Chih-Tang Sah, Jack S-C Sun and Joe J. Tzou, "Deactivation of boron acceptor in silicon by hydrogen," Appl. Phys. Letts. v43, 204-106, 15 July 1983. "Deactivation of group III acceptors in silicon during keV electron irradiation," Appl. Phys. Letts. v43(10), 962-964, 15 November 1983.

[931.2] Chih-Tang Sah, Sam S-C Pan and Charles C-H Hsu, "Hydrogenation annealing kinetics of group-III acceptors in oxidized silicon," J. Appl. Phys. v57(12), 5148-5161, 15 June 1985.

[931.3] Chih-Tang Sah, "Hydrogenation and dehydrogenation of shallow acceptors and donors in Si: fundamental phenomena and survey of literature," section 17.16 in Properties of Silicon, pp.584-604, INSPEC, the Institution of Electrical Engineers, London, New York, May 1988. ibid., et.al., "Hydrogenation and dehydrogenation of shallow acceptors and donors in Si: fundamental phenomena and survey of the literature,", section 17.17 on pp.604-612. For original figures and summary of literature data, see also the two SRC Annual Reports cited as references [13] for AR No.1 and [18] for AR No.2 in section 17.16 on p.600-601 Properties of Silicon.

[931.4] Chih-Tang Sah, "Studies of the Reliability Physics of Silicon VLSI Transistors," Annual Report No. 2 (01 August 1984 to 31 July 1985) of Contract No. SRC 83-01-030 and SRC Technical Report Series T85076, Semiconductor Research Corporation, P.O.Box 12053, Research Triangle Park, NC 27709 USA. 22pp. This report contains a detailed account of the discovery and followup experiments of hydrogenation and dehydrogenation of group-III acceptors in silicon. Some of which are also described and updated in the two sections in Properties of Silicon cited in reference [931.3].

[931.5] Chih-Tang Sah and S.C. Pan, "Summary of chronology of boron and group-III acceptor hydrogenation in crystalline silicon," Gordon Research Conference o Line Defects and Interfaces in Semiconductors, Plymouth State College, NH, USA. July 8-12, 1985.

[931.6] Chih-Tang Sah and S.C. Pan, "Experiments and model on deactivation of boron and other group-III acceptors at the Si-SiO_2 interface," Gordon Research Conference on Metal Insulator Semiconductor SYstems, Tilton School, Tilton, New Hampshire, USA. July 14-18, 1986.

999 PROBLEMS AND SOLUTIONS

P910.1 Derive the formula for the capture velocity of **(a)** electrons and **(b)** holes by interface traps. Let the volume capture rate of electrons and holes be given by c_n and c_o with unit (#-cm^3/s), and the volume density of the electron, hole, and occupied (by electron), unoccupied, and total occupied+unoccupied interface traps be given by n(x,y,z,t), p(x,y,z,t), n_T(x,y,z,t), p_T(x,y,z,t) and N_{TTv}(x,y,z) ≡ $n_T + p_T$, with unit (#/cm^3), where $N_{TTv}(x,y,z) \simeq N_{TT}(y,z)\delta(x)$ at the interface located at x=0 or $-x_1 < x < x_2$ around x=0.

P910.2 Derive the formula for the emission velocity of electrons and holes from interface traps. Let the volume emission rate of electrons and holes be given by e_n and e_p with unit (#/s), and the volume density of the occupied (by electron), unoccupied, and total occupied+unoccupied interface traps be given by n_T(x,y,z,t), p_T(x,y,z,t) and N_{TIv}(x,y,z) with unit (#/cm^3) where $N_{ITv}(x,y,z) \simeq N_{TI}(y,z)\delta(x)$ at the interface located at x=0 or $-x_1 < x < x_2$ around x=0.

P910.3 Derive the formula for the trapping velocity of **(1)** electrons and **(2)** holes for the cases of P910.1 and P910.2. (Hint, it is the capture minus the emission velocities and different for electrons and holes under arbitrary conditions.)

P910.4 Derive the steady-state electron-hole recombination velocity at the interface traps of P910.1-P910.3. Show that a simple expression is obtained for high concentration of electrons or holes at the interface.

P910.5 Show that if the DOS of the interface traps are distributed over the Si energy gap, then to a good approximation, the measured grtt and DOS is that averaged over about $2k_BT$ centered at the traps whose energy levels coincide with the Fermi energy at the interface. Consider the equilibrium case where $P(x=0,y,z) \times N(x=0,y,z) = n_i^2$.

P910.6 Show that the solution of P910.3 is more complicated at a nonequilibrium d.c. steady-state condition.

P910.7 Extend P910.4 to arbitrary time-dependent nonequilibrium conditions.

P911.1 Verify the numbers given in Table 911.1 and find the typographical errors in the table in Handler's original 1957 article (reference [30] in [1]) and in Table 5 of the most lucid review chapter by L. E. Katz in Chapter 4 on p.144 of VLSI TECHNOLOGY edited by S.M.Sze, McGraw-Hill Book Co. 1983 edition (reference [26] in [1]).

P911.2 The DOS of the surface traps on the clean (111) surface measured by the surface conductance experiments of Allen and Gobeli can also be fitted to $N_{ST}(V_S) = 9.5\times10^{13}\cosh(U_S+9.7)$ states/cm^{2}eV at 300K or k_BT=25.85meV. This, when integrated over the Si energy gap, must have a limited width so that the total surface state density does not exceed the silicon dangling bond density, 7.832×10^{14}cm^{-2} given in Table P911.1. Show that the separation of the two peaks is around $6k_{BT}$ to give a total trap number of 7.832×10^{14}cm^{-2} at each peak as anticipated by the silicon dangling bond density in the (111) plane given in Table P999.10A.

The total areal density of the surface states from silicon dangling bonds on a bare Si surface can be obtained by integrating the DOS, $N_{ST}(V_S)$, over the Si energy gap. Let T=300K and k_BT=0.2585eV, then we get

$$\begin{aligned}\int N_{ST}(V_S)dV_S &= \int 9.5\times10^{13}\cosh(U_S+9.7)dU_S\cdot k_BT \qquad \text{(P999.11A)}\\ &= 9.5\times10^{13}\times0.02585\times[\sinh(U_{Supper}+9.7) + \sinh(U_{Slower}+9.7)]\\ &= 9.5\times10^{13}\times0.02585\times[\sinh(U_{Supper}+9.7) + \sinh(U_{Slower}+9.7)]\\ &= 2.456\times10^{12}\times[\sinh(U_{Supper}+9.7) + \sinh(U_{Slower}+9.7)]\\ &\le \text{Silicon dangling bond density} = 7.832\times10^{14}\text{cm}^{-2}.\end{aligned}$$

We next assume that the DOS is symmetrical with respect to the neutral Fermi level,

$$\begin{aligned}E_{Fneutral} &= E_I - 9.7k_BT. \quad \text{Then,}\\ U_{Supper} &= -9.7 + \text{arcsinh}(7.832\times10^{14}/2.456\times10^{12})\\ &= -9.7 + 6.45 = -3.25 \Rightarrow -84\text{meV}\\ U_{Slower} &= -9.7 - \text{arcsinh}(7.832\times10^{14}/2.456\times10^{12})\\ &= -9.7 - 6.45 = -16.15 \Rightarrow -417\text{meV}\end{aligned}$$

Therefore, the width of the U-shaped DOS is about $2\times6.45k_BT$ = 333meV. A better representation would be a lower-donor and upper-acceptor DOS peak separated by about

300meV, centered at $E_I-9.7kT$, and each has a DOS of $7.8\times10^{14}cm^{-2}$. Both of these DOS's are from the amphoteric silicon dangling bond center, corresponding to its donor, neutral, and acceptor charge states, $Si^{\oplus}$, $Si\bullet^{\otimes}$ and $Si\bullet\bullet^{\ominus}$.

P911.3 Suppose that the bare, clean and ideal (not-reconstructed) silicon surface is exposed to atomic hydrogen in a VUV chamber at 300K and with a hydrogen pressure of 10^{-12} torr and no other gases in the vacuum chamber, and assume that each hydrogen atom will have a probability of 10^{-3} to stick to the Si surface (i.e. forming a H-Si bond). How long does it take to cover 99% of the dangling bonds if the clean Si surface is in the **(a)** (111), **(b)** (110) and **(c)** (100) directions? Do you expect a difference in time of coverage from the different crystallographic directions. (See Katz's review article cited as reference [26] in [911.1].)

P911.4 What is the areal surface charge density for each of the hydrogenated Si surfaces as a function of the surface potential bending U_S $(=q\psi_S/k_BT)$?

P911.5 Where is the equilibrium Fermi energy level at the surface if the Si is doped with 10^{18} cm^{-3} of **(a)** boron, **(b)** phosphorus, and **(c)** if the Si is pure and undoped?

P912.1 Show that the DOS resolution from the Terman method is about $4k_BT$. The Terman method, the DOS or D_{IT} versus E_T is obtained by comparing the experimental and theoretical HFCV curves of a MOSC (MOS capacitor) using the relationship between D_{IT} and $[(d/dV_G)(V_{G\text{-theory}} - V_{G\text{-experiment}})]\bullet(dV_G/dV_S)$ and the delta function approximation to the derivative of the Fermi-like trap occupation function, $df_T(E_T)/dE_T = (-1/k_BT)\bullet f_T(E_T)\bullet[1 - f_T(E_T)] \simeq \delta(E_T)$.

The solution was derived in FSSE and given by (683.9)

$$D_{IT}(E_T=E_F) = (C_o/q)(\partial\Delta V_G/\partial V_S) \quad \text{(states/eV-cm}^2\text{)} \qquad (683.9)$$

where

$$\Delta V_G = V_{G\text{-experiment}} - V_{G\text{-theory}}$$

the partial derivative is taken at constant $C_{g\text{-experiment}}$.

P912.2 Show that in the transition layer of the SiO_2/Si interface, if the Si or O bonds of the suboxides are ruptured then the partially bonded suboxide species may have eight single-dangling-bond configurations given by $Si_3O_y{\equiv}Si\bullet$ and $Si_3O_y{=}SiO\bullet$ where y=0,1,2 or 3; eight double-dangling bond configurations; and eight triple-dangling-bond configurations. Use the chemical symbol convention given here to list these configurations and also give several 2-D and 3-D sketches.

P912.3 Experimental DOS measurements using the Terman method on MOS capacitors were not available during the experiments which gave the suboxide concentrations in the SiO_2/Si transition layer listed Table 912.1. However, in view of the good correlation between the experimental suboxide concentrations and theoretical cut-bond densities, one can estimate the interface trap density assuming a certain percentage of the suboxide's Si do not have a chance to become bonded by oxygen during oxidation. **(a)** Take this percentage as 0.01% for a fully annealed state-of-the-art SiO_2/Si interface, and compute the total interface trap density in the entire Si energy gap in the SiO_2/Si layer. **(b)** Assume that the DOS of this Si-dangling bond interface trap can be approximated by that measured on the bare and clean Si surface by Allen and Gobeli given by (911.1). Give the equation of the DOS as a function of energy and Q_{IT} as a function of energy. **(c)** Assume that this is a donor-like interface trap located at E_I–0.4eV in the Si energy gap. What is its peak DOS if measured by the Terman method on a MOS capacitor at T=300K?

P913.1 Using the electron affinity and ionization potentials of an isolated atom in vacuum as given by Table 914.2, estimate the interface trap energy levels relative to the Si conduction band edge of the nitrogen dangling bond, N•, at the SiO_2/Si interface.

P913.2 There is a core difference between O^{+6} and N^{+5}. How does this difference affect the energy levels of the nitrogen dangling bond (Si-N•) estimated in P913.1?

P913.3 The column for F (fluorine) in Table 914.2 is missing. Look them up in the Handbook of Physics and Chemistry and fill up the column. Then, estimate the interface trap energy levels on a fluorine dangling bond, SiF•

P914.1 The kinetics of generation by the 10.2eV VUV light of the E_{MG} + 0.35 eV donor interface trap labeled P1 in Fig.914.1(a) follows the first order kinetics and the rate was measured (See Table 914.1.) [6], The details of the many microscopic-macroscopic steps are described by the chemical equations in (914.5z) (z=A,B,C,...) Write down the rate equations for each step and attempt to solve them simultaneously by picking out the dominant rate controlling step to give the zeroth solution, which is $[1-\exp(-\sigma F)]$ where σ is the generation cross section given by Table 914.1 and F is the fluence which could be the photon flux, photogeneration d.c. current through the MOSC, or the hole flux through through the oxide. **(a)** Find the first order solution by solving simultaneously the dominant and the next dominant rating limiting steps. **(b)** Is the generation cross section different if F is (i) the 10.2eV photon flux measured by a photon flux meter, (ii) the hole concentration generated at the gate/SiO_2 interface, (iii) the hole flux reaching the SiO_2/Si interface, (iv) the d.c. current flowing into the gate terminal, and (v) another flux you can think of.

P914.2 Obtain the detailed microscopic-macroscopic equations for generation of Si dangling bond interface traps by VUV dehydrogenation of the hydrogenated silicon dangling bonds analogous to (914.5x) (x=A,B,C,D,...). Note, the Si-H bond energy in vacuum is 3.10eV and the interband electron-hole pair generation threshold by an energetic electron is nearly equal to the silicon energy gap (E_G=1.12eV) as theoretically predicted by energy-momentum conservation theory by Sah (1968), and reanalyzed and experimentally verified by Yi Lu and Sah (See Physical Review B 52(8),5657,15-Aug-1995.)

P916.1 Obtain the bond-breaking distance for the potential distributions in the space-charge region of a reverse biased p/n junction with a reverse bias of V_R and a potential variation of $V(x) = Ax^n$ for the two hot carrier injection edges: **(a)** the low-field edge or x=0, and **(b)** the high-field edge or x=d where d is the SCR thickness. **(c)** Give physical realization or impurity profile of this potential function. (Not so easy)

The general relationships are first given below. The bond breaking distance for the two entry edges then follow.

$$V_{BB} = A d_{BB}{}^{n}$$
$$d_1 = [\ (V_{BB})/A]^{1/n}$$

$$V_D - V_{BB} = A d_{BB}{}^n$$
$$d_2 = [(V_D - V_{BB})/A]^{1/n}$$
$$V_D = A d^n$$
$$d = [(V_D)/A]^{1/n}$$

$$\text{(a) } d_{BB} = d_1 = [(V_{BB})^{1/n}]/A^{1/n}$$
$$\text{(b) } d_{BB} = d - d_2 = [(V_D)^{1/n} - (V_D - V_{BB})^{1/n}]/A^{1/n}$$

P916.2 Obtain the bond-breaking distance for the potential distributions in the space-charge region of a reverse biased n+/p junction with a reverse bias of V_R and an impurity profile of $N_{AA}(x\geq 0) = Ax^p$ and $N_{DD+}(x<0) = N_{DD+}$ = constant >> $N_A A(x>0)$ for the two hot carrier injection edges: **(a)** the high-field edge at x=0, and **(b)** the low-field or zero-field edge at x=d where d is the SCR thickness.

At $x=d$, $dV/dx=0$ and $V(d) = -V_D = -(V_R+V_{bi})$ where the built-in barrier height is $V_{bi} = (kT/q)\log_e[N_{DD+}N_{AA}(d_{bi})/n_i{}^2]$. The potential variation can be obtained from integrating the Poisson equation twice using these boundary conditions.

$$\varepsilon d^2V/dx^2 = -\rho = -(-q)N_{AA}(x) = qAx^p \quad \text{(P916.2A)}$$
$$dV/dx = [qA/\varepsilon(p+1)]\cdot(x^{p+1} - d^{p+1}) \quad \text{(P916.2B)}$$
$$V(x) = [qA/\varepsilon(p+1)]\cdot\{[x^{p+1}/(p+2)] - d^{p+1}\}\cdot x \quad \text{(P916.2C)}$$
$$V_{bi} = [qA/\varepsilon(p+2)]\cdot d_{bi}{}^{p+2} \quad \text{(P916.2D)}$$
$$= (k_BT/q)\log_e(N_{DD+}\cdot A d_{bi}{}^p/n_i^2) \quad \text{(P916.2E)}$$
$$V_R + V_{bi} = [qA/\varepsilon(p+2)]\cdot d^{p+2} \equiv V_D \quad \text{(P916.2F)}$$
$$\text{(a) } V_{BB} = -[qA/\varepsilon(p+1)]\cdot\{[d_{BB}{}^{p+1}/(p+2)] - d^{p+1}\}\cdot d_{BB} \quad \text{(P916.2G)}$$
$$\text{(b) } V_D - V_{BB} = -[qA/\varepsilon(p+1)]\cdot\{[(d-d_{BB})^{p+1}/(p+2)] - d^{p+1}\}\cdot(d-d_{BB}) \quad \text{(P916.2H)}$$

In both solutions (a) and (b) above, the functional dependence of d_{BB} or the bond-breaking efficiency, $\exp(-d_{BB}/\lambda_2)$, on the applied reverse bias voltage, V_R, can be obtained only iteratively for a given V_{BB}, N_{DD+}, and the parameters A and p for the $N_{AA}(x)$ distribution. The inverse problem of least-squares-fit (LSF) of the experimental data of the measured efficiency versus reverse-bias V_R to give the best LSF value of V_{BB} and λ_2 is more tedious but straight forward using the above solutions as parametric equations in the LSF analysis.

P921.01 Discuss the fundamental reasons of such a large difference between the two energies to release the electron trapped at the neutral oxygen vacancy, $V_O^{\ominus} \rightarrow V_O + e^{\ominus}$, measured by Scott Thompson [3] using tunneling emission (~1.0eV) and thermal emission (~0.4eV).

The energy measured by tunneling emission of the electron trapped at the neutral oxygen vacancy is the tunneling barrier height of the electron bound state in an electric field as depicted by Fig.141.5 on p.45 of FSSE for a Coulomb barrier of a hydrogen atom which shows large barrier lowering (peak is lowered significantly below the vacuum level) by the applied electric field due to the large range of the $1/r^2$ Coulomb force. However, the electron trapped to the neutral oxygen vacancy is bound to a short-range neutral potential well of $V_O^{\ominus}$, rather than the long-range Coulomb potential well. Thus, the tunneling barrier height is nearly equal to that of the bound state energy, $E_C - E_A$, with only a small field-induced barrier lowering whose magnitude depends on the radius, size or range of the short-range neutral potential well. The logarithmic tunneling rate is inversely proportional to height and the thickness of the tunneling barrier while the thickness is reduced by increasing the applied electric field. Thus, the slope of the electron tunneling emission rate versus electric field gives the tunneling barrier height or the quantum mechanical bound state energy level or binding energy of the electron trapped at the neutral oxygen vacancy center.

The thermal activation energy measured from the temperature dependence of the thermal emission rate of the electron bound to the neutral oxygen vacancy is a measure of the energy change of the entire system consisting of the oxygen vacancy and its surrounding Si and O atoms. In this case, the system energy difference is that of the SiO_2 with and without an electron trapped at a neutral oxygen vacancy. This energy difference includes the energy from lattice relaxation, that is, the change of the potential energy of the silicon and oxygen atoms surrounding the bridging-oxygen-vacancy when they move (or relax) back to their original locations after the trapped electron is completely removed from the oxygen vacancy (moved to infinity). This position relaxation gives off some energy because the silicon atoms without the trapped electron are more positively charged and hence are pushed away from each other further than when the electron is trapped, i.e. the Si-Si bond length at the V_O is longer without the trapped electron. Thus, when the electron trapped at E_C-E_A=1.0eV is (thermally) released by absorbing the thermal vibration energy of the silicon and oxygen atoms, the lattice relaxation of the surrounding silicon and oxygen atoms to their original positions of a neutral oxygen vacancy would give off some energy, in this case, about 0.6eV, making the observed thermal activation energy or the total energy change, 1.0eV – 0.6eV ≈ 0.4eV. This is known as the Jahn-Teller effect.

The atomic configuration difference is demonstrated pictorially by the two atomic configurations given in Fig.921.1(a) for $V_O^{\ominus}$ and Fig.921.1(b) for $V_O^{\otimes}$. Note that the Si:Si bond length I drew for $V_O^{\ominus}$ of figure (a) is shorter than that in $V_O^{\otimes}$ of figure (b) because the trapped electron in $V_O^{\ominus}$ reduces further the repulsive Coulomb force between the two Si^{+4} cores adjacent to the oxygen vacancy.

The final atomic configurations of these two charge states are easier to compute from quantum mechanics than the energy difference and the transition rates of electron emission by thermal emission, e_{n-1}^{t}, by tunneling emission, ω_{n-1}, and of electron capture by thermal relaxation of the lattice, c_{n0}^{t}. A multiphonon configuration coordinate model was developed by Prof. Kun Huang [11] which has been widely used to treat the large Jahn-Teller shift between the thermal activation and optical energies and the large temperature dependence of the thermal capture rates observed in compound semiconductors and ionic insulators, and also some impurity traps in Si by Alex Wang [12] and other elemental semiconductors. This large energy shift in SiO_2 observed by Thompson [3] was a historical first, nevertheless, he was also able to measure the rates quite accurately and over a wide range of conditions (temperature and electric field), and his experimental data are sufficient to provide predictive design of transistor reliability.

P921.02 Is the optical excitation energy of the electron trapped at the neutral oxygen vacancy similar to the tunneling emission energy or the thermal activation energy described in P921.01 and why?

The optical excitation energy is the photon energy necessary to release a trapped electron from the negative oxygen vacancy which returns it to the neutral charge state, $V_O^{\ominus} + h\nu \rightarrow V_O^{\otimes} + e^{\ominus}$. In the experiment, the time dependence of the charged oxygen vacancy density is measured by the Voltage axis shift of the MOS C_g-V_G or MOST I_D-V_G curve. Thus, the optical excitation produced total trapped charge change and its derivative with respect to photon energy, would not involve lattice relaxation energy, and hence the optical DOS versus energy is similar to that obtained by the electric-field dependence of tunneling emission.

P921.03 Give the reasons on why the optical emission of the electrons trapped at the neutral oxygen vacancy electrons has not been reported?

Because of the low thermal activation energy, the optical experiments must be carried out at low temperatures so that the electrons trapped at the neutral oxygen acceptor will not be thermally excited out before the optical excitation experiment can be carried out. The experimental difficulty is compounded by the necessity of accurate measurement of the trapped electron charge and its change due to optical emission at the low temperatures, such as 77K, and the need of an optically transparent gate conductor electrode on the oxide film around 1eV photon energy which is harder to fabricate reliably and reproducibly compared with a non-transparent gate conductor electrode. In principle, the optical emission should be observable at low temperatures, however, special transistor structures and techniques of electron injection into the oxide and capture by the oxide electron trap have not been available to one experimenter.

P921.04 What is the cause of the broad density-of-states of the trapped electrons to the neutral oxygen vacancy observed by the tunnel emission measurement?

This is a reflection of the random distribution of the silicon oxygen bond length (Si-O-Si) of the amorphous SiO_2 film grown thermally on a crystalline Si substrate which would in turn affect the Si-Si bond length around the oxygen vacancy. Thus, the DOS (Density of States) measured from the tunneling emission spectroscopy provides an experimental measure on the crystallinity of the SiO_2 film grown on crystalline Si substrate. If an atomic epitaxial growth can be developed to grow SiO_2 film on crystalline Si substrate, then it is conceivable that not only the broad density-of-states would narrow to a sharp line but its amplitude would be drastically decreased due to lower oxygen vacancy density. Even in thermally grown film not under epitaxial grown conditions, one would expect the initial atomic layers of SiO_2 film to approach the crystallinity of its seed, the silicon substrate. Thus, the broadness and the amplitude should both decrease with decreasing oxide film thickness. These thickness dependencies are not expected for the 14.6-28.1nm range (See Fig.11 of [3].) and have not been observed in the 2-5nm range since the trapped electrons tend to tunnel out of the oxygen vacancy when the film is less than about 8.6nm (86 Angstroms) as indicated in Fig.P921.04 given below. Thus, the injected and captured electrons at the oxygen vacancy, shown in Fig.P921.1(b), cannot be retained for a sufficiently long time to carry out the tunneling experiments even at low temperatures. It may be possible to carry out the hole tunneling experiment for the deeper hole trap at $(E_D-E_V)_{th} \simeq 1.44\text{eV}$.

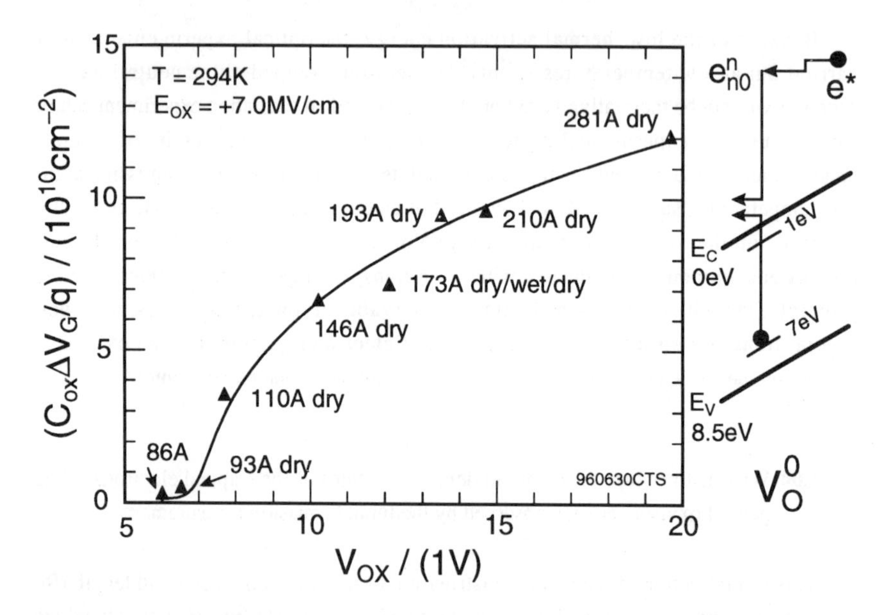

Fig.P921.04 Steady-state oxide electron charge density, $Q_{OT}/q = -C_{OX}\Delta V_G/q$, as a function of oxide voltage drop, V_{OX}, after an electron injection from Si of $N_{INJ}=1.0\times10^{16}\text{cm}^{-2}$ at $E_{OX}=+7.0\text{MV/cm}$, for $X_{OX} = 8.6\text{nm}$ to 28.1nm . Data from Fig.14 of Thompson [3]. Inset shows electron-impact emission of the bond electron from the neutral oxygen vacancy, $V_O^{\otimes} + e^{-*}(>7\text{eV}) \rightarrow V_O^{\oplus} + 2e^-$.

P921.05 Discuss the fundamental reasons that tunneling emission of trapped holes was not observed from the density-of-states of the hole trap due to the positively charged oxygen vacancy measured by Yi Lu [4] using thermal-stimulated hole emission shown in Fig.P921.2.

This thermal-stimulated hole emission measurement, just like the oxide electron trap discussed in P921.01, gives the thermal activation energy which contains a Jahn-Teller lattice relaxation contribution. From the atomic configuration of $V_O^{\otimes}$ and $V_O^{\oplus}$ shown in Fig.P921.1(c) and (d) respectively, it is evident that thermal hole emission $V_O^{\oplus} \rightarrow V_O^{\otimes}+$

$h^{\oplus}$ could give off a very large lattice relaxation energy because the two Si-O bonds are now longer which can overcompensate for the additional bond electron in the Si-Si. Thus, the binding energy of the hole could be much larger than the 1.44eV, two or three times, which would place it in the energy range near the energy gap of silicon. On the other hand, if the bond energy change of the two Si-O bonds and the Si-Si bond are similar during reconfiguration between the two charge states, $V_O^{\otimes}$ and $V_O^{\oplus}$, then the thermal activation energy and the hole tunneling emission energy would be similar. Thus far, in preliminary one-shot experiments, it had not been possible to tunnel out the trapped holes even in thin (5.5nm) oxides at large electric fields [4].

P921.06 The density of the discrete hole traps at E_V+1.44eV decreases with decreasing oxide thickness while that of the distributed hole traps in E_V+0.15eV to E_V+1.35eV appeared independent of oxide thickness in the range of 5.5nm to 11nm. Explain these in terms of the possibility of crystallinity and randomness of the SiO_2 lattice.

The first few layers of oxide films, thermally grown on a crystalline substrate, may retain some of the crystallinity property of the substrate which is randomized in the region away from the SiO_2/Si in the thicker SiO_2 films. Since the rather narrow 1.44eV±0.07eV level is associated with oxygen vacancy, it is consistent with the lower oxygen vacancy concentration in the first few layers of SiO_2 because the initial oxide growth is limited by Si-O reaction rather oxygen diffusion. In thicker films, the oxide growth at the SiO_2/Si interface is limited by oxygen diffusion through the already grown SiO_2 film, thus, oxygen deficiency is expected to be more prevalent or the density of bridging oxygen vacancy should be higher as observed.

On the other hand, the distributed U-shaped DOS was attributed to random Si-O bond length and angle of the amorphous SiO_2 whose randomness reaches a constant average as the film becomes thicker as indicated by Yi Lu's Fig.4 [4]. Thinner SiO_2 films (<5.5nm) were not available at the time of his experiments (1994), however, one would expect a trend of lower density in thinner SiO_2 films similar to that of the 1.44eV oxide hole trap whose areal density or total number appears to decrease towards zero at zero oxide thickness as expected from seed induced crystallinity for the first few atomic layers of the oxide film grown on a silicon crystalline substrate.

P921.07 Elaborate on the possible fundamental reasons concerning the sharpness of the DOS of oxide hole trap in comparison with the broadness of the DOS of the oxide electron trap shown in Fig.P921.2.

The broadness of the DOS of the oxide electron traps was attributed to the random Si-O bond and angle which produce the large random Si-Si bond length at the V_O center which binds the electron. The sharper DOS of the oxide hole trap indicates the expected lesser dependence on the randomness of the Si-O bond length and angle because the holes are trapped at the O_{2p} orbital which is much more tightly bounded and hence perturbed less by the surrounding random Si-O bonds. This is consistent with a smaller lattice relaxation shift or smaller Jahn-Teller effect.

P921.08 In the above problems, we tacitly assumed that the oxygen vacancies are distributed in a thin plane sheet in the SiO_2. Experimental evidences indicated that there was a spatial distribution peaked around 8nm from the SiO_2/Si interface. For example, the centroid, X_{OT}, measured from the SiO_2/Si interface, of the electron-impact generated positive oxide charge, i.e. V_O^+, reported by Thompson showed that $X_O - X_{OT} \simeq 85\pm5$A or the oxide trap is located about 8nm from the gate-conductor(polysilicon)/SiO_2 interface for X_{OX}=146A to 210A at E_{OX}=7MV/cm (See Fig.11 of [3].) which was further collaborated with its density which showed a threshold at about 86A (See Fig.14 of [3].) A further evidence of spatial or depth distributions was indicated by the dependence of X_{OT} on the oxide electric field or the kinetic energy of the injected electron (Fig.12 of [3].), following $X_{OT} \simeq (E_C - E_D)/E_{OX}$ giving $X_{OT} \bullet E_{OX} \simeq (E_C - E_D)$ = 7.9eV, 7.7eV, 7.5eV for E_{OX} = 8MV/cm, 7MV/cm and 6MV/cm. This spatial distribution was sketched in Fig.681.1 on p.646 of FSSE where X_{OT} was measured from the gate/oxide interface rather than the oxide/Si interface designated by Thompson. Discuss the effect of spatial distribution on the measured energy distribution of the oxide trap shown in Fig.P921.2.

The rather wide energy distribution of the oxide electron trap at the oxygen vacancy center shown in Fig.P921.2, with a DOS peak at $E_C - E_A \simeq 1.0$eV and FWHM$\simeq$0.5eV (Full Width at Half Maximum), measured by Thompson using the tunnel emission of trapped electron method, has been a continued suspect of a spatial distribution in disguise since the random Si-O bond length and angle may not be able to account for so large a FWHM. This uncertainty was resolved, to a large extend, by Yi Lu's DOS measurements of the oxide hole trap at the oxygen vacancy center shown in Fig.921.2, using the thermally stimulated emission method, which showed a rather narrow peak, FWHM≈0.07eV. These two sets of measured 'DOS' would be consistent with a spatial distribution whose centroid is about 80A from the gate/oxide interface, if it were not for Yi Lu's DOS measured in 55A oxide with a total oxide hole trap density of $5.5\times10^{13}(cm^{-2}eV^{-1})\times0.07eV = 3.8\times10^{12}cm^{-2}$ while Thompson's total oxide electron trap density in the 150A and 281A oxides was about $3.5\times10^{12}\times0.5 = 1.8\times10^{12}cm^{-2}$. We note that tunnel emission of trapped electrons measures the oxide traps in a thin slice, Δx, because the tunneling rate is exponentially dependent on tunneling distance while the trapped holes in the entire oxide were emitted or detrapped by thermal emission. Thus, the two sets of data are consistent with oxygen vacancies spatially distributed throughout the thickness of the oxide film, whose energy distribution is fairly narrow and sharp, ≈0.07eV via thermally stimulated hole emission measured by Yi Lu. Thus, Thompson's broad tunnel DOS peak is then due to from the spatial distribution.

P921.09 The characterization of the oxide electron trap ($E_C-E_A = 1.0eV$) and oxide hole trap ($E_C-E_A = 7eV$) by Thompson [3] was applied by Kavalieros [13] to characterize the oxide-charge-scattering-limited electron mobility in the silicon surface channel of nMOST by controlled generation of positive oxide charges. When the voltage drop through the oxide film, V_{OX}, is less than $(E_C-E_A)/q=7V$, the neutral oxygen vacancy can only be charged negatively by the injected electron as indicated by Fig.P921.09(b) (the lower figure) on the following page. When V_{OX} is greater than 7V, electron impact emission of one of the two bond electrons in the neutral oxygen vacancy will dominate to charge the neutral oxygen vacancy positively as indicated by Fig.P921.09(a) (the upper figure). A bipolar-MOS transistor structure is used to inject the electrons from the silicon substrate into the oxide which is shown in Fig.P921.09(c) on the page following the next. The following problems can be worked out from the positive and negative charging data shown in Fig.P921.09(a) and (b).

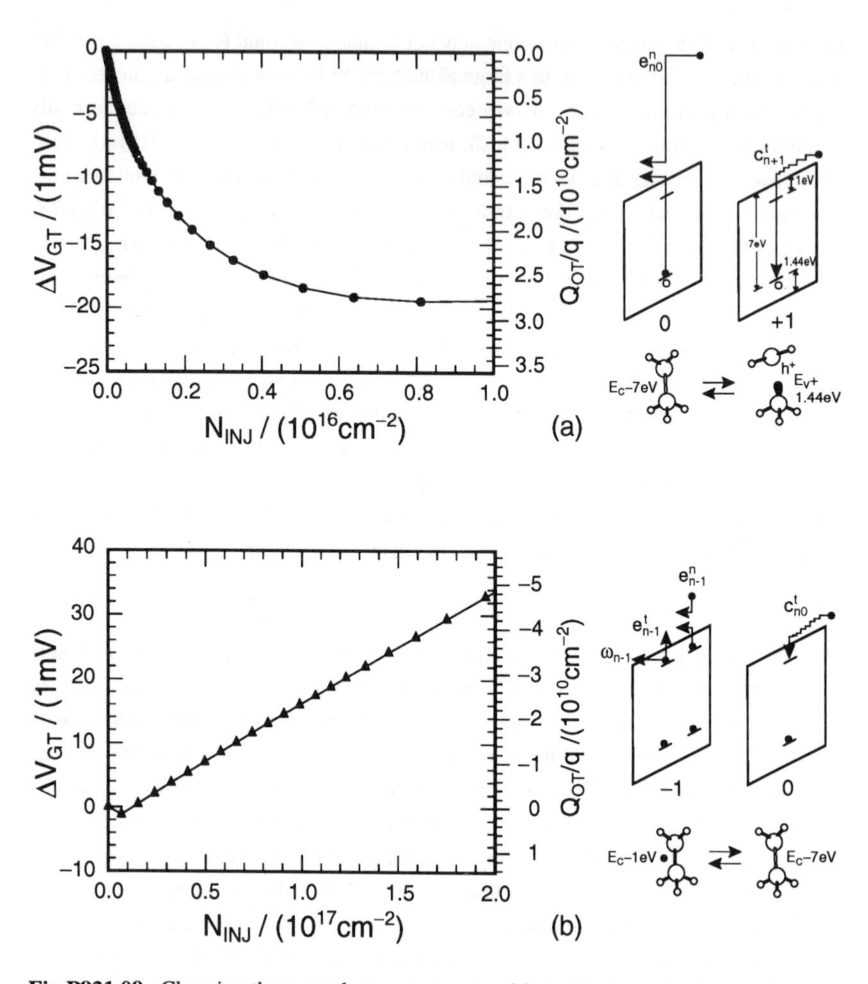

Fig.P921.09 Charging the neutral oxygen vacancy **(a)** positively via impact emission of bond electron by injected and accelerated energetic electrons, and **(b)** negatively via thermal capture of injected electron. The electrons are injected into the SiO_2 from the silicon substrate using the substrate hot electron injection method (SHEi) in a npnBJT/nMOST or nBiMOS transistor structure shown in **(c)**. From [13].

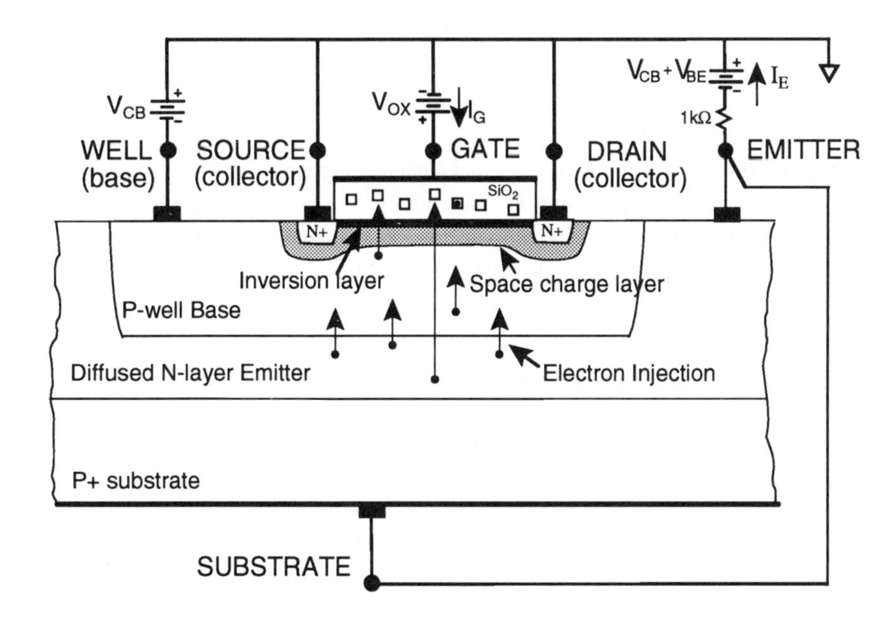

Fig.P921.09(c) nBiMOS transistor structure used for substrate hot electron injection (SHEi) to charge the neutral oxygen vacancy positively or negatively.

(a) What is the oxide thickness?

Use the y-axes of Fig.P921.09(a), and $C_{ox}\Delta V_{GT} = -Q_{OT}$ then

$$X_{OX} = \varepsilon_{ox}/C_{ox} = (\varepsilon_{ox}/q)[\Delta V_{GT}/(-Q_{OT}/q)]$$
$$= [(3.9\times 8.854\times 10^{-14}F/cm)/1.602\times 10^{-19}C]$$
$$\times[-20\times 10^{-3}/(-2.9\times 10^{10}cm^{-2})] = 148A$$

(b) What is the electron capture efficiency by the neutral oxygen vacancy?

We assume that the curve in Fig.P921.09(b) is the initial linear rise of a simple one-exponential, as inferred from the impact emission curve in Fig.P921.09(a). Then, using $N_{OT}(N_{INJ}) = Q_{OT}(N_{INJ})/q = (Q_{OT-\infty}/q)[1 - \exp(-\sigma_{n0}^{t}N_{INJ})]$, the efficiency is defined as the initial slope of the N_{OT} vs N_{INJ} plot, namely, the number of negative oxide charge generated for each electron injected, or the fraction of the injected electrons which are captured by the neutral oxygen vacancy. This is given by

$$\eta_0 \equiv [d/d(qN_{INJ})]Q_{OT}(N_{INJ}\rightarrow 0) = \sigma^t_{n0}\cdot(Q_{OT\infty}/q)$$
$$= 5\times10^{10}cm^{-2}/2.0\times10^{17}cm^{-2} = 2.5\times10^{-7}.$$

This is an extremely small efficiency: about 4 out of 10^6 injected electrons are captured and most of the injected electrons will pass through the oxide. In order to obtain the fundamental parameter, the capture cross-section, we must know the total neutral oxygen vacancy, $N_{OT\infty}$, which is not provided by Fig.P921.09(b).

(c) What is the impact efficiency and cross-section of emission of the bond electron in neutral oxygen vacancy from an energetic electron impact?

Using the decay (or rise) curve in Fig.P921.09(a) for electron impact, we get

$$\Delta V_{GT}(N_{INJ}\rightarrow\infty)=-19.5mV\ .$$

By setting $\sigma^n_{n0}N_{INJ}=1$ at $\Delta V_{GT}(N_{INJ})/\Delta V_{GT\infty}=0.632$, in

$$N_{OT}(N_{INJ}) = Q_{OT}(N_{INJ})/q = (Q_{OT\infty}/q)[1 - \exp(-\sigma^n_{n0}N_{INJ})]$$

we get

$$\Delta V_{GT} = 0.632\times\Delta V_{GT\infty} = 0.632\times(-19.5mV) = -12.3mV.$$

We then read off from Fig.P921.09(a), and get

$$N_{INJ} = 0.175\times10^{16}cm^{-2}.$$

Thus, the electron-impact cross-section for emission of a bond electron at the neutral oxygen vacancy is

$$\sigma^n_{n0} = 1/0.175\times10^{16}cm^{-2} = 5.7A^2.$$

The impact efficiency is the initial slope and also the product of the cross-section and the final steady-state value. It can be readily be obtained from Fig.P921.09(a) to give

$$\eta_0 = \sigma^n_{n0}\cdot N_{OT\infty} = 2.8\times10^{10}/0.175\times10^{16} = 16\times10^{-6}.$$

This is much larger (160/2.5≈64 times larger) than the cross section of thermal capture of electrons. Part of the reason is that during thermal capture, there are simultaneous electron emission by thermal and tunneling pathways as indicated by the left transition energy band diagram in Fig.P921.09(b). These emission transitions make the measured capture cross section smaller. We would also expect a strong electric field dependence from tunneling, and a large temperature dependence from thermal emission, both of which were observed and characterized by Thompson [3].

P921.10 In view of the possibility of both energy and space distribution of the oxygen vacancy as described in the previous problem, what are some of the experiments necessary to separate out these two distributions?

P931.1 Discuss the reasons why the various rejected atomic configurations of the hydrogenated boron acceptor in silicon shown in Figs.931.2 and 931.3 are less favorable energetically than those given in Figs.931.1(a),(b),(c). Based on simple considerations of lattice displacement or reconstruction energy and the four equivalent positions of H^+, show that Fig.931.1(c) is probably the lower energy configuration than Figs.931.1(a) and (b).

P931.2 Based on the simple bond energy consideration, show that hydrogenation of group-V donors in Si is unlikely because it is unstable and dissociates at room temperature. Discuss situations in which hydrogen could be bound to a group-V donor and show that the configuration is likely to be one in which the hydrogen is not at the B-Si bond center position or even inside the bond.

The hydrogen could form a side or wing terminator bond with the extra or fifth valence electron of the group-V donor, but this must be at one of the wing positions and the two-electron hydrogen terminator bond is likely to be weak due to the +5 charge of the donor core (compared with the +3 charge of the acceptor core). To have this extra electron attached to the donor, the temperature must be lowered to about 4K so that the phosphorus donor is deionized or having a Si conduction electron trapped to it. So, the experimental condition to provide the hydrogen is limited to exposure to hydrogen plasma on the surface or x-ray and keV-electrons since to release the hydrogen from the gate or nearby sources by energetic electrons using electrical bias would be difficult if not more tedious due to freeze out and inability of creating a Si region or surface space-charge-layer of sufficiently high electric field or voltage drop. The chemical formula would look like

```
        H
       /
  Si3≡P-Si≡Si3
```

P932.1 The chemical reaction equations given in the text do not show the core charge of each element nor the charge of the electron represented by the dot and two electrons represented by the dash to simplify the symbol and reduce the symbol length. This is compounded by not showing all the valence electrons of each atom either, for example, the six valence electrons in the oxygen atom, because that would require putting dots above and below the element symbol, preventing the representation of the atomic cluster of a complex center to be written on one line in the word-processor representation. Thus, this does not meet the 'one-glance' criteria for charge balance since the core charge number must be memorized because not all the valence electrons in each atom are shown, only those in the bonds, hence one cannot use the total number of electrons shown in the bond plus the missing bond electrons to arrive at the core charge number. Illustrate these deficiencies by writing out the core charges in (932.1).

$$(Si^{+4})_3{\equiv}(Si^{+4}){\bullet}(H^{+1}){\bullet}(B^{+3}){\equiv}(Si^{+4})_3 + \text{Heat}$$
$$\rightarrow (Si^{+4})_3{\equiv}(Si^{+4})\text{-}(B^{+3}){\equiv}(Si^{+4})_3 + h^+ + (H^+){\bullet} \qquad (932.1A)$$

P932.2 Give the detailed steps or chain of the acceptor hydrogenation reaction represented by only one chemical reaction given by (932.7) in the MOS structure. The hydrogen comes from the 'dissolved' or weakly bonded hydrogen in the gate and gate/oxide interface (either the aluminum, refractory metal, refractory metal silicide, or polycrystalline silicon gate) which is freed or released (or emitted from the hydrogen trap) by several possible excitations: hot electrons from avalanche hot electron injection, substrate hot electron injection, or Auger electron-hole recombination in n+Si, tunnel electron injection which is accelerated by the oxide electric field, and x-ray/keV-electron ionizing radiation. (See equations (3) to (5b) in reference [930.2]).

P932.3 Give the differential equation for each chain step in P932.2.

The bonded or bound hydrogen at the hydrogen source is freed, released or emitted from the hydrogen or proton trap by an applied excitation other than heat since heat will also dissociate the hydrogen from the A•H•Si complex. We represent the excitation by e* which is an energetic electron with kinetic energy higher than the bond energy of the weakly bonded hydrogen in the hydrogen source whose binding energy (probably ~ <0.5eV as suggested by electromigration and other experiments) is less if not much less than the binding energy at the A•H•Si (~>1eV to ~3eV). If the excitation has higher energies, then e* is the decayed product of the higher energy primary or secondary electrons or holes, via electron-electron collision and plasmon decay. The single-body arrow, $\rightarrow$, represents reaction (grt or generation, recombination, trapping). The double-body arrow, $\Rightarrow$, represents transport via diffusion and drift. The reaction-transport equations are listed as follows for the MOS capacitor on a boron-doped p-Si substrate. The corresponding differential equations are also given. The hydrogen or proton trap at the hydrogen source, the gate region, is labeled by subscript T which has a total hydrogen-trap volume concentration of $M_{TT}(x,y,z)$ cm^{-3} and a hydrogenated trap volume concentration of $m_T(x,y,z,t)$. The interface traps at the SiO_2/Si interface, both acting as protonic and electronic traps, are represented by $m_{IT}(t)$ cm^{-2}, however, the time constants of protonic and electronic trapping differ by orders of magnitude, so we use different letter symbols, m_T or m_{IT} for protonic traps and n_I or n_{IT} for electronic traps or electron traps since we have a symbol left for hole traps, p_T and p_{IT}, while proton-hole or anti-proton does not exist in our situation (or energy range or electronic material which would exist in protonic materials in which the entire solid is constructed by protonic bonds or hydrogen bridges, such as Si•H•Si•••$(H\bullet Si)_3$ and other elements.

Si•H + e*	$\rightarrow$	Si• + H•↑	(gate)	(P932.2A)
H• + H•	$\rightarrow$	H_2	(gate)	(P932.2B)
H•↑ (gate)	$\Rightarrow$	H•	(diffuses through SiO_2 and ignored oxide protonic traps)	(P932.2C)
H• + Si•	$\rightarrow$	Si-H	(at SiO_2/Si)	(P932.2D)
H• (SiO_2/Si)	$\Rightarrow$	H•	(diffuses through surface space-charge layer)	(P932.2E)
H• + Si≡B-Si	$\rightarrow$	Si≡B•H•Si + e^-		(P932.2F)
e^- + h^+	$\rightarrow$	perfect	(electron-hole recombine)	(P932.2G)
H• + H•	$\rightarrow$	H_2	(hydrogen recombination)	(P932.2H)

$$(\partial H/\partial t)_A = e_H^* m_T - c_H H(M_{TT} - m_T) \quad (P932.3A)$$
$$(\partial H/\partial t)_B = g_H H_2 - r_H H^2 \quad (P932.3B)$$
$$(\partial H/\partial t)_{Gate} = (\partial H/\partial t)_A + (\partial H/\partial t)_B \quad (at\ x=-x_{ox}) \quad (P932.3AB)$$
$$\partial H/\partial t = D_H \partial^2 H/\partial x^2 \quad (P932.3C)$$
$$D_{Si}(\partial H/\partial x) - D_H(\partial H/\partial x) = -(\partial m_{IT}/\partial t)_D \quad (P932.3D)$$
$$(\partial m_{IT}/\partial t)_t = c_{Hi} H(M_{ITT} - m_{IT}) - e_{Hi} m_{IT}) \quad (P932.3E)$$
$$\partial H/\partial t = D_{Si} \partial^2 H/\partial x^2 \quad (P932.3F)$$
$$(\partial m_T/\partial t)_t = c_{Hv} H(M_{TT} - m_T) - e_{Hv} m_T \quad (P932.3G)$$
$$(\partial H/\partial t)_v = g_{Hv} H_2 - r_{Hv} H^2 \quad (P932.3H)$$

P941.1 The traditional electromigration equation given by (940.1) was derived empirically by fitting experimental data and using the ad hoc assumption of a drift or electron-wind force from electron-atom momentum transfer. We now give, for the first time, the analytical solutions of the new void-surface bond-breaking model, proposed by Sah in the early 1980's and described in the text, using the fundamental linear electron impact model for breaking the metallic electron bond that binds the metal atom to the void's surface.

The model used in this analysis is shown in Fig.P941.1. The random shape of the void depends of the void's seed. It is approximated by a square. The areal density of the electron current tangential to the void surface or the Void/Metal interface is scaled by just the remaining line width, $W_W - w_V(t)$, thus $J(t) = J_0\{1-[w_V(t)/W_W]\}^{-1}$ which will be assumed constant over the remaining line width. The density symbols are as follows. M_{TT} = total number of surface metal atom site, occupied plus unoccupied by the metal atom. m_T = surface site occupied by metal atom or trapped metal atom (adatom). m = mobile or activated metal atom whose diffusivity is D_m and whose thermal activation energy is E_A. There are two limiting solutions. **(a)** Metal-bond-breaking rate limited. **(b)** Diffusion limited by diffusion of activated metal atom in the metal line, either via the vacancy mechanism (self-diffusion) or along the grain boundaries (surface diffusion). The rate-transport equations are listed below. In the flux equation, (P941.1), **we drop the atomic drift current term**, $\mu_m E_x m$, which is the term all previous momentum-transfer/wind-force models have originated using a current-proportional fudge factor.

$$F_M = -D_m \partial m/\partial x + \mu_m E_x m \simeq -D_m \partial m/\partial x \qquad (P941.1)$$

$$\partial m/\partial t = -\partial F_M/\partial x + e_M m_T - c_M(M_{TT} - m_T)m \qquad (P941.2)$$

$$= +D_m \partial^2 m/\partial x^2 + (e_M + c_M m)m_T - c_M M_{TT} \qquad (P941.3)$$

$$\partial m_T/\partial t = -(e_M + c_M m)m_T + c_M M_{TT} \qquad (P941.4)$$

$$e_M = e_{M1}J \quad \text{or} \quad e_M = e_{M2}J^2 \qquad (P941.5)$$

$$J = J_0/[1-(w_V/W_W)] \equiv J_0/(1-\omega) \quad \text{where } \omega \equiv w_V/W_W \qquad (P941.6)$$

$$R/R_0 = 1/[1-(w_V/W_W)] \equiv 1/(1-\omega) \qquad (P941.7)$$

$$(R/R_0)-1 \equiv \Delta R/R_0 = \omega/(1-\omega) \qquad (P941.8)$$

The release or emission rate of the metal atoms trapped on the void's surface, m_T, is proportional to the electron current density, $e_M = e_{M1}J$, near the surface because the drifting electrons break the metallic bond. This activates the trapped metal atom, m_T, from its surface trapping site, M_{TT}. At higher current densities we have a J^2 dependence, $e_M = e_{M2}J^2$, because the drifting electrons also enhance vacancy migration to provide more vacant sites for the released atoms to move into.

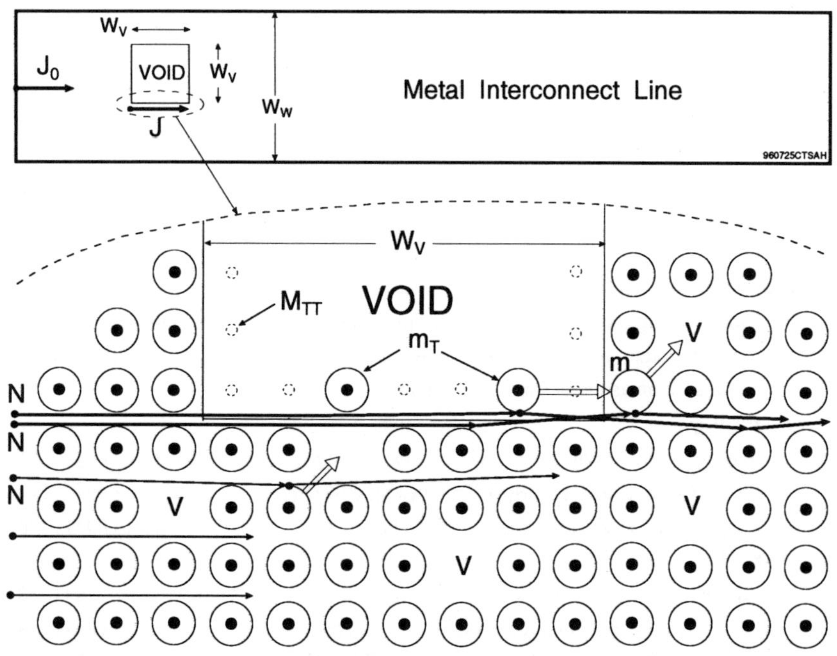

Fig.P941.1 An ideal square void model for electromigration in a metal interconnect line. The void in the upper figure is enlarged in the lower figure which is approximately to scale. Small dot = electron. Solid arrow = electron trajectory. Small broken circle = surface site for Al atom. Circled dot = Al^{+3}= 0.26Å. Large circle = Al valence electron orbit = 1.044Å-1.312Å (radii from FSSE Table 162.2). $\lambda_{ele\text{-}mfp}$ = 150Å. λ_{Debye} = 0.576Å. $d_{Al\text{-}Al}$ = 2.86Å. Open-arrow = Al track = inverse track of Al-vacancy.

(a) Void-Surface-Bond-Breaking Rate Limited

$$D_m\partial^2 m/\partial x^2 << (e_M + c_M m)m_T - c_M M_{TT} \quad (P941.9)$$

$$\partial w_V/\partial t = k\partial m/\partial t = k(e_M + c_M m)m_T - kc_M m M_{TT} \quad (P941.10)$$

$$\simeq k(e_M + \quad)m_T \quad (P941.10A)$$

$$= k(e_{M1}J)\bullet[M_{TT}\exp(-E_A/k_BT) \quad (P941.10B)$$

$$= k(e_{M1}J_0)[M_{TT}\exp(-E_A/k_BT)]/[1-(w_V/W_W)] \quad (P941.10C)$$

$$\partial(w_V/W_W)/\partial t = k(e_{M1}J_0/W_W)M_{TT}\exp(-E_A/k_BT)/[1-(w_V/W_W)] \quad (P941.11A)$$

$$\equiv \partial\omega/\partial t = (1/\tau)/(1-\omega) \quad (P941.11B)$$

$$\int dt/\tau = t/\tau = \int[1-(w_V/W_W)]d(w_V/W_W) \equiv \int(1-\omega)d\omega \quad (P941.12A)$$

$$= \{1 - [1-(w_V/W_W)]^2\}/2 = \omega(1-\omega/2) \quad (P941.12B)$$

$$1/\tau = k(e_{M1}M_{TT}/W_W)J_0\exp(-E_A/k_BT) \quad \text{(P941.12C)}$$

$$(R/R_0) - 1 = \omega/(1-\omega) \quad \text{(P941.13)}$$

$$t_\infty/\tau = 1/2 \quad \text{Breakdown Time at } R=\infty \quad \text{(P941.14A)}$$

$$(R/R_0) - 1 = \Delta R/R_0 \simeq t/\tau \quad \text{Short Time} \quad \text{(P941.14B)}$$

$$TTF = t(\Delta R/R_0=0.1) = \tau/10$$

$$= (W_W/10ke_{M1}M_{TT})\bullet(J_0)^{-1}\bullet\exp(+E_A/k_BT) \quad \text{(P941.15)}$$

If $e_M=e_{M2}J^2$, then $TTF \propto (J_0)^{-2}$. (P941.15A)

(b) Bulk Diffusion Limited

Surrounding the void, m_T release or m generation reaction is much faster than m diffusing, therefore, m and m_T at the void's surface, x=0, has reached the steady-state as soon as the current is turned on.

At x = 0

$$e_Mm_T - c_Mm(M_{TT}-m_T) = 0 \quad \text{with } m_T(x=0,t)=M_{T0},\ m(x=0,t)=M_0 \quad \text{(P941.16A)}$$

$$M_0 = e_MM_{T0}/[c_M(M_{TT} - M_{T0})] \simeq (e_M/c_M)(M_{T0}/M_{TT}) \quad \text{(P941.17A)}$$

$$= (e_{M1}J_0/c_M)[1/(1-\omega)]\exp(-E_A/k_BT) \quad \text{(P941.17B)}$$

At x > 0

$$\partial m/\partial t = D_m\partial^2m/\partial x^2 \quad \text{(P941.18)}$$

$$m(x,t) = M_0\text{erfc}[x/2\surd(D_mt)] \quad \text{(P941.19)}$$

$$F_m(x=0,t) = -D_m\partial m/\partial x|_{x=0} = D_mM_0/(\pi D_mt)^{1/2} \quad \text{(P941.20)}$$

$$Q_m(x=0,t) = \int F(x=0,t)dt = 2M_0(D_mt/\pi)^{1/2} \quad \text{(P941.21)}$$

Let $w_V(t) = kQ_m(x=0,t)$ (P941.22A)

$$D_m = D_{m0}\exp(-E_{Dif}/k_BT) \quad \text{Vacancy or Grain Boundary} \quad \text{(P941.22B)}$$

$$M_{T0} = M_{TT}\exp(-E_A/kT) \quad \text{Thermally activated surface atoms.} \quad \text{(P41.22C)}$$

$$d(w_V/W_W)/dt = (d/dt)(kQ_m/W_W) \equiv d\omega/dt = (k/W_W)F_m(x=0,t) \quad \text{(P941.23)}$$

$$\int(1-\omega)d\omega = \omega(1-\omega/2) \equiv \surd(t/\tau) \quad \text{(P941.23A)}$$

$$= 2k(e_{M1}J_0/c_MW_W)\exp[-(E_A+E_{Dif}/2)/k_BT](D_{m0}/\pi)^{1/2}\int dt/\surd t \quad \text{(P941.23B)}$$

$$= 2k(e_{M1}J_0/c_MW_W)\exp[-(E_A+E_{Dif}/2)/k_BT](D_{m0}/\pi)^{1/2}2\surd t \quad \text{(P941.23C)}$$

$$\tau \equiv (\pi/D_{m0})(c_MW_W/4ke_{m1})^2J_0^{-2}\exp[-(2E_A+E_{Dif})/k_{BT}] \quad \text{(P941.24)}$$

$$t = [\omega(1-\omega/2)]^2\tau \quad \text{(P941.25)}$$

$$\Delta R/R_0 = \omega/(1-\omega)$$

$$t_\infty = t(\omega=1) = \tau/4 \quad \text{Breakdown Time at } R=\infty \quad \text{(P941.25A)}$$

$$\Delta R/R_0 \simeq \omega = \surd(t/\tau) \quad \text{Short Time} \quad \text{(P941.25B)}$$

$$TTF = t(\Delta R/R_0=0.1) = \tau/100$$

$$= TTF_0\bullet(J_0)^{-2}\bullet\exp[+(2E_A+E_{Dif})/k_BT] \quad \text{(P941.26)}$$

If $e_M=e_{M2}J^2$, then $TTF \propto (J_0)^{-4}$. (P941.26A)

950 Transistor Reliability Problems and Solutions

P950.1 Calculate the operation time–to–failure (TTF_{OP}) of a state-of-the-art MOS transistor using the physics of charging oxide traps, the Coulombic scattering of electrons by the oxide charge, and the MOS transistor's linear and subthreshold characteristics. The following parameters are given.

x_{OX} = 100 Å
$N_{OT\infty}$ = 10^{11} cm^{-2} (Increase this to 10^{12} cm^{-2} if MOST does not fail)
σ_n = 10^{-16} cm^2 (Effective cross section of charging oxide traps)
μ_0 = 600 $cm^2/V-s$ (Before oxide traps are charged)
μ_{OX} = 6000 $cm^2/V-s$ (Charged oxide traps are near SiO_2/Si interface)
μ_{OX} =60000 $cm^2/V-s$ (Charged oxide traps are away from interface)

Calculate the TTF using the two criteria (a) and (b) and indicate how $J_{G\text{-stress}}(t)$ is determined at each operating voltage.

(a) $\Delta V_T = 20$ mV at operating gate voltages of 5, 3.3, and 2.5V.

(b) $\Delta I_D/I_D = 2\%$ at operating gate voltages of 5, 3.3, and 2.5V.

Assume that the oxide-trap-charging pathway starts by Fowler–Nordheim tunneling of Si conduction electrons (either from Si–gate or Si–substrate) and that σ_n is the value at V_G=5V. To get σ_n at 3.3V and 2.5V, you need to take into account the dependence on oxide electric field or oxide voltage drop.

This problem may be solved at several levels of device–physics sophistication. The lowest level, practiced by most factory engineers, is to assume that $V_{OX}=V_G$ and $E_{OX}=|V_G/x_{OX}|$. The answers may be off by several orders of magnitude. To give more accurate solutions one must assume a substrate dopant impurity concentration, a specific channel type, nMOST or pMOST, and a V_G bias polarity. Thus, the problem is open–ended whose solutions require increasing sophistication in device physics and give increasing accuracy; aside from the more difficult material physics on the transition rate of each step of the oxide charging pathway whose voltage, electric field, and temperature dependencies must be taken into account to give the most complete result.

Solutions:

The nMOST is inverted since $V_G > V_{GT}$. The zeroth approximation for the oxide voltage is $V_O = V_G$, however, this is inaccurate for calculation of the Fowler–Nordheim tunneling current which is an exponential function of the oxide electric field. Therefore, solve for V_S and use V_S to calculate V_O in the equation for V_G. Assume $N_{AA}=2\text{x}10^{17}$ cm^{-3} and T=300K.

P950.1(a) $\Delta V_T = 20$mV at $V_{GG}=V_{DD}=5.0$V, 3.3V, 2.5V.

$V_{AA} = \varepsilon_s q N_{AA}/(2C_o^2)$ [FSSE-SG p.127, and FSSE p.658] (682.8B)

$C_o = \varepsilon_o/x_0 = 3.9\text{x}8.854\text{x}10^{-14}(\text{F/cm})/100\text{x}10^{-8}\text{cm} = 3.453060\text{x}10^{-7}\text{F/cm}^2$

$V_{AA} = (11.9\text{x}8.854\text{x}10^{-14}\text{F/cm})(1.602\text{x}10^{-19}\text{C})(2\text{x}10^{17}\text{cm}^{-3})$

$\times[2(3.453060\text{x}10^{-17}\text{F/cm}^2)^2]^{-1} = 0.140\text{V}$

$V_S = 2V_F + (kT/q)\log_e\{[(V_G-V_{FB}-V_S)^2/((4kT/q)V_{AA}\Delta_I)]\}$ FSSE-SG(411.16I)

$V_{FB} = \phi_{MS} - (Q_{OT} + Q_{IT})/C_o$ FSSE-SG(410.32)

Assume $N_{GG}=5\text{x}10^{19}\text{cm}^{-3}$ and n+G/SiO_2/p–Si nMOST.

$\phi_{GX} = \phi_G - \phi_X$

$= (V_{FG} - V_{FX}) + (\phi_{Ge} - \phi_{Xe})$

$= -((kT/q)\log_e(N_{GG}/n_i) + (kT/q)\log_e(P_{XX}/n_i))$

$= -(kT/q)\log_e(N_{GG}P_{XX}/n_i^2)$

$= -(0.0259\text{V})\log_e((5\text{x}10^{19})(2\text{x}10^{17})/(10^{10})^2) = -1.014\text{V}$

$V_{FB} = \phi_{GX} - (Q_{OT} + Q_{IT})/C_o = \phi_{GX} = -1.014\text{V}$ when $Q_{OT} = Q_{IT} = 0.0$

$V_F = V_{FX} = (kT/q)\log_e(P_{XX}/n_i) = (0.0259\text{V})\log_e(2\text{x}10^{17}/10^{10}) = 0.435\text{V}$

Consider V_G=5V

Let $V_S^{j=0} = 2V_F = 0.8708$V, and use $V_S^{j=0} \equiv V_S^0$ where j=iteration number.

$\Delta_I^0 = 33.62248$ where $U_B = V_S^0/(kT/q) = 33.622 = 2U_F$

$V_S^1 = 2V_F + (kT/q)\log_e\{[(V_G-V_{FB}-V_S^0)^2/(4(kT/q)V_{AA}\Delta_I^0)]\} = 0.870$

$+ (0.0259)\log_e\{[5-(-1.014)-0.8708]^2/(4\text{x}0.0259\text{x}0.140\text{x}33.6)\}$

$= 0.9734$ V

$\text{Error}^1 = V_S^1 - V_S^0 = 0.9734 - 0.8708 = 0.1026$V

$\Delta_I^1 = 1.0000$

$V_S^2 = 0.870 + (0.0259)\log_e\{[5-(-1.014)-0.9734]^2/(4\text{x}0.0259\text{x}0.140\text{x}1)\}$

$= 1.063$ V

$\text{Error}^2 = V_S^2 - V_S^1 = 1.063 - 0.9734 = 0.0896$V

$\Delta_I^2 = 1.0000$

$V_S^3 = 0.870 + (0.0259)\log_e\{[5-(-1.014)-1.063]^2/(4\text{x}0.0259\text{x}0.140\text{x}1)\}$

$= 1.063$ V

$\text{Error}^3 = V_S^3 - V_S^2 = 1.063 - 1.063 = 0.000$V **Converged.**

$V_0 = V_G - V_{FB} - V_S = 5 - (-1.014) - 1.063 = 4.951$V

The gate tunnel current is given by:

$J_G = J_N\langle 100\rangle = 2.6\text{x}10^6 E_0^2 \exp(-238.5/E_0)$ FSSE (385.1A)

$E_0 = V_0/x_0 = 4.951\text{V}/(100\text{x}10^{-8}\text{ cm}) = 4.951$ MV/cm

$J_G = (2.6x10^6 A/cm^2)(cm/MV)^2(4.951MV/cm)^2 exp(-238.5MV/cm/4.951MV/cm)$
$= 7.647x10^{-14}$ A/cm^2

$\Delta V_G = -(\Delta Q_{OT}(t) + \Delta Q_{IT}(t))/C_o = -\Delta Q_{OT}(t)/C_o$ where $\Delta_{QIT}(t)=0$ assumed.

$\Delta V_G = (qN_{OT\infty}/C_o)[1 - e^{-t/\tau}]$ where $\tau = q/(J_{Gstress}\sigma_n(E_0))$

Given $\sigma_n(E_0) = 10^{-16}$ cm^2 (value at V_G=5V, V_0=4.951V).

$\Delta V_G(TTF) = (qN_{OT\infty}/C_o)[1 - exp(-TTF/\tau)]$

$exp(-TTF/\tau) = 1 - C_o\Delta V_G(TTF)/(qN_{OT\infty})$

$TTF = -\tau log_e[1 - C_o\Delta V_G(TTF)/(qN_{OT\infty})]$

$\tau = q/[J_{Gstress}\sigma_n(E_0)]$
$= (1.602x10^{-19}C/(7.647x10^{-14}A/cm^2x10^{-16}cm^2))$
$= 2.094954x10^{10}s$

(a.1) $V_{DD}=V_{GG}$=5V: At $N_{OT\infty} = 10^{11}cm^{-2}$

$TTF = -\tau log_e[1 - C_o\Delta V_G(TTF)/(qN_{OT\infty})]$
$= -(2.094954x10^{10}s)log_e[1-(3.453x10^{-7}F/cm^2)(20x10^{-3}V)$
$/(1.602x10^{-19}C\times10^{11}cm^{-2})]$
$= 1.18x10^{10}$ s = 374 years

(a.2) $V_{DD}=V_{GG}$=5V: At $N_{OT\infty} = 10^{12}cm^{-2}$

TTF = $9.23x10^8$ s = 29 years

Consider V_G=3.3V:

$V_S^0 = 2V_F = 0.8708V$

$\Delta_I^0 = 33.62248$ where $U_\theta = V_S^0/(kT/q) = 33.622 = 2U_F$

$V_S^1 = 2V_F + (kT/q)log_e\{[(V_G-V_{FB}-V_S^0)^2/(4(kT/q)V_{AA}\Delta_I^0)]\} = 0.870$
$+ (0.0259)log_e\{[3.3-(-1.014)-0.8708]^2/(4x0.0259x0.140x33.6)\}$
$= 0.9526$ V

$Error^1 = V_S^1 - V_S^0 = 0.9526 - 0.8708 = 0.0818V$

$\Delta_I^1 = 1.0000$

$V_S^2 = 0.870 + (0.0259)log_e\{[3.3-(-1.014)-0.9526]^2/(4x0.0259x0.140x1)\}$
$= 1.0424$ V

$Error^2 = V_S^2 - V_S^1 = 1.0424 - 0.9526 = 0.0898V$

$\Delta_I^2 = 1.0000$

$V_S^3 = 0.870 + (0.0259)log_e\{[3.3-(-1.014)-1.0424]^2/(4x0.0259x0.140x1)\}$
$= 1.0410$ V

$Error^3 = V_S^3 - V_S^2 = 1.0410 - 1.0424 = -0.0014V$

$\Delta_I^3 = 1.0000$

$V_S^4 = 0.870 + (0.0259)\log_e\{[3.3-(-1.014)-1.0410]^2/(4\times0.0259\times0.140\times1)\}$
$= 1.0410$ V

$\text{Error}^4 = V_S^4 - V_S^3 = 1.0410 - 1.0410 = 0.000\text{V}$ **Converged.**

$V_0 = V_G - V_{FB} - V_S = 3.3 - (-1.014) - 1.0410 = 3.273\text{V}$

$E_0 = V_0/x_0 = 3.273\text{V}/100\times10^{-8}\text{cm} = 3.273$ MV/cm

Following the calculation steps for the 5V case,

$J_G = 6.2759\times10^{-25}$ A/cm^2 (Tunnel current.)

(a.1) $V_{GG}=V_{DD}=3.3$V: At $N_{OT\infty} = 10^{11}\text{cm}^{-2}$

$TTF = -\tau\log_e[1 - C_o\Delta V_G(TTF)/(qN_{OT\infty})]$

$\tau = q/(J_{Gstress}\sigma_n(E_0))$

The electron capture cross section is given at V_G=5V where V_O=4.951 V and E_O=4.951 MV/cm. At V_G=3.3V, V_O=3.273V, and E_O=3.273MV/cm, the capture cross section is given by

$$\sigma_n(E_0') \simeq \sigma_n(E_0)(E_0/E_0')^n$$

The capture of electron by coulombic trap is determined by the interaction of the electron with the oxide trap through the Coulomb force which is inversely proportional to the kinetic energy of the electron. Thompson experimentally measured the oxide field dependence of the capture cross section and showed n=2–4 [J. Appl. Phys., 72, pp. 4683–4695, 1992]. Let n=3 then

$\sigma_n(E_0'=3.273\text{ MV/cm}) \simeq \sigma_n(E_0=4.951\text{MV/cm})\cdot(4.951/3.273)^3$
$= (10^{-16}\text{ cm}^2)(4.951/3.273)^3$
$= 3.5\times10^{-16}\text{ cm}^2$

$\tau = (1.602\times10^{-19}\text{ C})/[(6.2759\times10^{-25}\text{A/cm}^2)(3.5\times10^{-16}\text{cm}^2)]$
$= 7.293\times10^{20}$ s

$TTF = -(7.293\times10^{20}\text{ s})\log_e[1-C_o\Delta V_G(TTF)/(qN_{OT\infty})]$
$= 4.11\times10^{20}$s
$= 1.3\times10^{13}$years **(Will not fail!)**

(a.2) $V_{GG}=V_{DD}=3.3$V: At $N_{OT\infty} = 10^{12}\text{cm}^{-2}$

$TTF = 3.21\times10^{19}$ s
$= 1.0\times10^{12}$ years **(Will not fail!)**

Consider V_G=2.5V:

$V_S^0 = 2V_F = 0.8708\text{V}$

$\Delta_I^0 = 33.62248$ where $U_\theta = V_S^0/(kT/q) = 33.622 = 2U_F$

$V_S^1 = 2V_F + (kT/q)\log_e\{[(V_G-V_{FB}-V_S^0)^2/(4(kT/q)V_{AA}\Delta_I^0)]\} = 0.870 +$
$+ (0.0259)\log_e\{[2.5-(-1.014)-0.8708]^2/(4\times0.0259\times0.140\times33.6)\}$

$= 0.9390\text{V}$

$\text{Error}^1 = V_S^1 - V_S^0 = 0.9390 - 0.8708 = 0.0682\text{V}$

$\Delta_I^1 = 1.0000$

$V_S^2 = 0.870 + (0.0259)\log_e\{[2.5-(-1.014)-0.9390]^2/(4\text{x}0.0259\text{x}0.140\text{x}1)\}$

$= 1.0286\text{V}$

$\text{Error}^2 = V_S^2 - V_S^1 = 1.0286 - 0.9390 = 0.0896\text{V}$

$\Delta_I^2 = 1.0000$

$V_S^3 = 0.870 + (0.0259)\log_e\{[2.5-(-1.014)-1.0286]^2/(4\text{x}0.0259\text{x}0.140\text{x}1)\}$

$= 1.0268\text{V}$

$\text{Error}^3 = V_S^3 - V_S^2 = 1.0268 - 1.0286 = -0.0018\text{V}$

$\Delta_I^3 = 1.0000$

$V_S^4 = 0.870 + (0.0259)\log_e\{[2.5-(-1.014)-1.0268]^2/(4\text{x}0.0259\text{x}0.140\text{x}1)\}$

$= 1.0268\ \text{V}$

$\text{Error}^4 = V_S^4 - V_S^3 = 1.0268 - 1.0268 = 0.000\text{V}$ **Converged.**

$V_0 = V_G - V_{FB} - V_S = 2.5 - (-1.014) - 1.0268 = 2.487\text{V}$

$E_0 = V_0/x_0 = 2.487\text{V}/100\text{x}10^{-8}\text{cm} = 2.487\ \text{MV/cm}$

Following the calculation steps of the 3.3V case,

$J_G = 3.637\text{x}10^{-35}\ \text{A/cm}^2$ (Tunnel current.)

$\sigma_n(E_0') \simeq \sigma_n(E_0)(E_0/E_0')^n$

Let n=3

$\sigma_n(E_0'=2.487\ \text{MV/cm}) \simeq \sigma_n(E_0=4.951\ \text{MV/cm})(4.951/2.487)^3$

$= (10^{-16}\ \text{cm}^2)(4.951/2.487)^3$

$= 7.9\text{x}10^{-16}\ \text{cm}^2$

$\tau = q/[J_{Gstress}\sigma_n(E_0)]$

$= (1.602\text{x}10^{-19}\text{C})/[(3.637\text{x}10^{-35}\text{A/cm}^2)(7.9\text{x}10^{-16}\ \text{cm}^2)]$

$= 5.575\text{x}10^{30}\text{s}$

(a.1) $V_{DD}=V_{GG}=2.5$V: At $N_{0T\infty}=10^{11}\text{cm}^{-2}$

$\text{TTF} = -(5.575\text{x}10^{30}\ \text{s})\log_e[1-C_o\Delta V_G(\text{TTF})/(qN_{0T\infty})]$

$= 3.14\text{x}10^{30}\text{s}$

$= 1.0\text{x}10^{23}$years **(Will not fail.)**

(a.2) $V_{DD}=V_{GG}=2.5$V: At $N_{0T\infty}=10^{12}\text{cm}^{-2}$

$\text{TTF} = 2.46\text{x}10^{29}\ \text{s}$

$= 7.8\text{x}10^{21}$ years **(Will not fail.)**

P950.1(b) $\Delta I_{D\text{-sat}}/I_{D\text{-sat}} = 2\%$ at $V_{GG}=V_{DD}$=5.0V, 3.3V, 2.5V.

$$I_{D\text{-sat}} = (Z/L)(\mu C_0/2)\cdot(V_G-V_T)^2$$
$$\Delta I_{D\text{-sat}}/I_{D\text{-sat}} = [(Z/L)(\mu' C_0/2)(V_G-V_T')^2 - (Z/L)(\mu_0 C_0/2)(V_G-V_{T0})^2]\times$$
$$\times[(Z/L)(\mu_0 C_0/2)(V_G-V_{T0})^2]^{-1}$$
$$= [\mu'(V_G-V_T')^2 - \mu_0(V_G-V_{T0})^2]/[\mu_0(V_G-V_{T0})^2]$$
$$= \{[\mu'(V_G-V_T')^2]/[\mu_0(V_G-V_{T0})^2]\} - 1$$

where

$$1/\mu' = 1/\mu_L + [N_{OT\infty}/n_{OT}(t)]^{-1}(1/\mu_{ox})$$
$$= 1/\mu_0 + \{N_{OT\infty}/N_{OT\infty}[1 - \exp(-t/\tau)]\}^{-1}(1/\mu_{OX})$$
$$1/\mu' = 1/\mu_0 + [1 - \exp(-t/\tau)](1/\mu_{ox})$$
$$\mu_0/\mu' = 1 + \mu_0[1 - \exp(-t/\tau)](1/\mu_{ox})$$
$$\mu'/\mu_0 = \{1 + [1 - \exp(-t/\tau)](\mu_0/\mu_{ox})\}^{-1}$$
$$\Delta I_{D\text{-sat}}/I_{D\text{-sat}} = (\mu'/\mu_0)[(V_G-V_T')/(V_G-V_{T0})] - 1$$
$$= \{1+ [1-(\mu_0/\mu_{ox})\exp(-t/\tau)]\}^{-1}[(V_G-V_T')/(V_G-V_{T0})]^2-1$$

Approximate $[(V_G-V_T')/(V_G-V_{T0})]^2 \simeq 1$ to get an analytical solution which gives the (μ'/μ_0) dependence for the $\Delta I_{D\text{-sat}}/I_{D\text{-sat}}$. If the $((V_G-V_T')/(V_G-V_{T0}))^2$ is retained, approximations can be made to give an analytical solution which shows that the μ'/μ_0 contribution is larger than $[(V_G-V_T')/(V_G-V_{T0})]^2$ contribution for the give μ_{ox} numbers.

$$\Delta I_{D\text{-sat}}/I_{D\text{-sat}} \simeq \{1 + (\mu_0/\mu_{ox})[1 - \exp(-t/\tau)]\}^{-1} - 1$$
$$1 + (\mu_0/\mu_{ox})[1 - \exp(-TTF/\tau)] = [1/(1-0.02)] = 1/0.98$$

Substituting $\Delta I_{D\text{-sat}}/I_{D\text{-sat}} = -0.02$.

$$TTF = -\tau\cdot\log_e[1 - (0.02/0.98)(\mu_{ox}/\mu_0)]$$

(b.1) $V_{DD}=V_{GG}$=5V:

From previous analysis in (a.1) at $V_{GG}=V_{DD}$=5V,

V_0 = 4.951 V
E_0 = 4.951 MV/cm
J_G = 7.647×10^{-14} A/cm^2
$\sigma_n(E_0)$ = 10^{-16} cm^2
τ = 2.0949×10^{10} s

Given μ_{ox} = 6000 cm^2/V-s ($\pm Q_{OT}$ near SiO_2/Si interface.)
μ_0 = 600 cm^2/V-s
TTF = $-(2.0949\times10^{10}s)\log_e[1 - (0.02/0.98)(6000/600)]$
= 4.782×10^9 s = 129 years

Given μ_{ox} = 1000 cm^2/V-s
TTF = 7.249×10^8 s = 23 years.

(b.1) $V_{DD}=V_{GG}=3.3V$:

From previous analysis in (a.1) at $V_{GG}=V_{DD}=3.3V$,

$V_0 = 3.273$ V

$E_0 = 3.273$ MV/cm

$J_G = 6.276\times10^{-25}$ A/cm^2

$\sigma_n(E_0) = 3.5\times10^{-16}$ cm^2

$\tau = 7.293\times10^{20}$ s

Given $\mu_{ox} = 6000$ cm^2/V-s ($\pm Q_{OT}$ near SiO_2/Si interface.)

$\mu_0 = 600$ cm^2/V-s

TTF $= -(7.293\times10^{20}\text{s})\log_e[1 - (0.02/0.98)(6000/600)]$

$= 1.665\times10^{20}$ s $= 4.487\times10^{12}$ year

(b.1) $V_{DD}=V_{GG}=2.5V$:

From previous analysis in (a.1) at $V_{GG}=V_{DD}=2.5V$,

$V_0 = 2.487$ V

$E_0 = 2.487$ MV/cm

$J_G = 3.637\times10^{-35}$ A/cm^2

$\sigma_n(E_0) = 7.9\times10^{-16}$ cm^2

$\tau = 5.575\times10^{30}$ s

Given $\mu_{ox} = 6000$ cm^2/V-s ($\pm Q_{OT}$ near SiO_2/Si interface.)

$\mu_0 = 600$ cm^2/V-s

TTF $= -(5.575\times10^{30}\text{ s})\log_e[1 - (0.02/0.98)(6000/600)]$

$= 1.273\times10^{30}$ s $= 3.470\times10^{22}$ years.

P950.2 Calculate the operation time–to–failure (TTF_{op}) of a state–of–the–art BiCMOS silicon bipolar junction transistor (BJT) using the physics of charging oxide traps and generation of the interface traps, the SNS BJT theory and the partition method. The following parameters are given for the n+/p/n BJT.

$X_{OX} = 100$ Å

$Y_B = 1.00$ μm

$W_E = 1.00$ μm = Does not matter. Will be normalized out.

$W_C = 2.00$ μm = Assume one–sided base contact.

Z = Does not matter. Will be normalized out.

$X_B = 0.10$ μm

$N_{EE} = 10^{20}$ cm^{-3} (Disregard degeneracy and E_G narrowing.)

μ = Find all mobility values in FSSE. Do not use values

from other source books, so we have one numerical answer from all of you.

P_{BB} = 10^{18} cm^{-3}
N_{CC} = 10^{17} cm^{-3}
$N_{OT\infty}$ = 10^{12} cm^{-3} (Increase to 10^{12} cm^{-3} if BJT does not fail.)
σ_n = 10^{-16} cm^2 (Effective cross section of charging n_{OT}.)
μ_0 = 600 cm^2/V-s (Before oxide traps are charged.)
μ_{OX} = 6000 cm^2/V-s (Charged $\pm Q_{OT}$ near SiO_2/Si interface.)
μ_{OX} = 60000 cm^2/V-s (Charged $\pm Q_{OT}$ away from interface.)
N_{IT0} = 0 (Assume a simple symmetrical interface trap.)
$N_{IT\infty}$ = 10^{12} trap/cm^2 **(Increase to 10^{13} cm^{-2} if BJT does not fail.)**
E_T = E_I at the midgap, and
c_{ns} = c_{ps} = $\sigma_s\theta_s$ = $10^{-16}\times10^7$ = 10^{-9} cm^3/s or
S_{n0} = S_{p0} = 10^3 cm/s or 10^4 cm/s.

The electron–channel on the p–base will be assumed to be terminated by an open circuit approximated by an infinitely long transmission line since at the defined failure condition, the channel is probably not too strongly inverted.

Calculate the TTF at V_{EB}(reverse) = +5V, +3.3V, +2.5V using the criteria $\Delta h_{FE}/h_{FE}$ = 20% at a normal collector current for this geometry. Indicate how $J_{G\text{-}stress}(t)$ is determined at each operating voltage. (Hint, pre-stress h_{FE} should be > 20, and you may use a higher value.)

Assume that the pathway for oxide trap charging starts by (1) Fowler–Nordheim tunneling of Si conduction electrons from the n–inversion layer on the p–Si base and (2) avalanche injection of electrons from the surface depletion layer on the n–channel on the p–Si base. Assume that the values of the cross-sections of positive oxide trap charging and interface trap generation are those at V_{EB}=+5V. Calculate σ_n at 3.3V and 2.5V, using their dependence on the oxide electric field or oxide voltage drop.

This problem can be solved at several levels of device–physics sophistication. The simplest level, practiced by most factory engineers, is to assume that V_{OX}=V_{EB}=+5V and E_{OX}=$|V_G/x_{OX}|$=5MV/cm. The answers may be off by several orders of magnitude. The problem is open–ended whose solutions require increasing sophistication in device physics and give increasing accuracy. aside from the more difficult material physics on the transition rate of each step of the oxide charging pathway whose voltage, electric field, and temperature dependencies must be taken into account to give the most complete result.

Solution: As derived in C-T Sah's lectures.

$$-I_C = -A_E(qN_BD_B/X_B)[\exp(qV_{EB}/kT)-1]\alpha_{BF}$$
$$-(A_E+A_{OL})qX_{CB}[(e_ne_p)/(e_n+e_p)]N_{TT}$$
$$-I_{CH}\gamma_{E-ch}\alpha_{BF-ch} \text{ where } \alpha_{BF} = \text{sech}(X_B/L_B) \simeq 1 - X_B^2/(2D_B\tau_B)$$
$$-I_E = +A_E\{[qN_BD_B/X_B + qP_ED_E/X_E'(s)][\exp(qV_{EB}/kT)-1]$$
$$+(qn_iX_{EB}/\tau_{EB})[\exp(qV_{EB}/2kT)-1]\} + I_{CH}$$

By KCL - Kirchoff's Current Law,

$$I_B = -I_C - I_E$$
$$= -A_E(qN_BD_B/X_B)[\exp(qV_{EB}/kT)-1][1 - X_B^2/(2D_B\tau_B)]$$
$$-I_{CBO} - I_{CH}\gamma_{E-ch}\alpha_{BF-ch}$$
$$+A_E(qN_BD_B/X_B)[\exp(qV_{EB}/kT)-1]$$
$$+A_E[qP_ED_E/X_E'(s)][(\exp(qV_{EB}/kT)-1]$$
$$+A_E(qn_iX_{EB}/\tau_{EB})[\exp(qV_{EB}/2kT)-1] + I_{CH}$$
$$I_B = A_E\{[(qN_BD_B/X_B)X_B^2/(2D_B\tau_B) + (qP_ED_E/X_E')]\times[\exp(qV_{EB}/kT) - 1]$$
$$+(qn_iX_{EB}/\tau_{EB})[\exp(qV_{EB}/2kT)-1]\}$$
$$-I_{CBO}$$
$$+I_{CH}(1 - \gamma_{E-ch}\alpha_{BF-ch})$$

If the emitter thickness, $X_E'(s)$, is thin compared to the minority carrier diffusion length, then the minority carriers (holes) will reach the n+poly/n+emitter interface where they recombine at the specified surface recombination velocity, S_{E0}.

$$I_B = +A_E\{[(qN_BD_B/X_B)(X_B^2/2D_B\tau_B) + (qS_{EO}P_E)][\exp(qV_{EB}/kT)-1]$$
$$+(qn_iX_{EB}/\tau_{EB})[\exp(qV_{EB}/2kT)-1]\}$$
$$-I_{CBO} + I_{CH}(1 - \gamma_{E-ch}\alpha_{BF-ch})$$

When the BJT is stressed by reverse–bias of the emitter/base junction, the change of the base current is given by:

$$\Delta I_B(\text{stress}) = (qA_E\Delta S_EP_E)[\exp(qV_{EB}/kT) - 1] + \Delta I_{CH}(1 - \gamma_{E-ch}\alpha_{BF-ch})$$

This gives a degradation of the two–port forward hybrid current gain or common–emitter current gain, $h_{FE} = (I_C - I_{CBO})/I_B$.

$$\Delta h_{FE} = (\partial h_{FE}/\partial I_B)\Delta I_B = -[(I_C - I_{CBO})/I_B^2]\Delta I_B$$
$$\Delta h_{FE}/h_{FE} = -(\Delta I_B/I_B)$$

Next, we relate ΔS_E and ΔI_{CH} to the interface trap generation mechanism.

$$S_E = S_p = c_{pe}N_{IT}$$
$$\Delta S_E = c_{pe}\Delta N_{IT} = c_{pe}N_{IT\infty}[1 - \exp(-t/\tau)] \text{ assuming } N_{ITO} = 0.$$

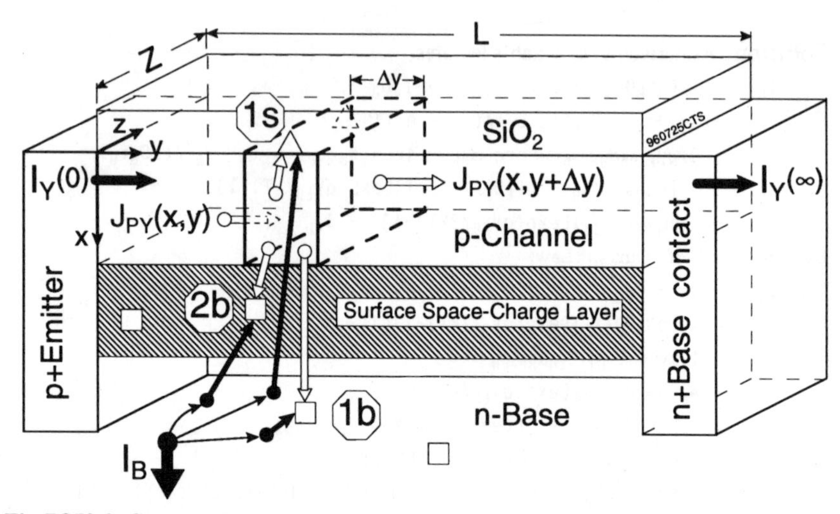

Fig.P950.2 Cross-sectional view and current components of a hole surface channel connected to the p+emitter under the n-base oxide in a Si p+/n/p BJT. For n+/p/n BJTs, interchange n and p. Rectangle: bulk traps, Triangle: interface traps.

Consider the channel current in Fig.P950.2. [Sah, "A new semiconductor tetrode - the surface–potential controlled transistor," Proc.IEEE,49,pp.1623–1634,Nov.1961.]

$$I_Y = Z\int_0^\infty J_{NY}(x,y)dx$$

At steady–state, $\partial J_{NY}/\partial y - qR_N = 0$ (P950.2A)

where $R_N = R_N(1b) + R_N(1s) + R_N(2b)$ and

$$R_N(1b) = [(PN - n_i^2)N_{TT}c_nc_p]/[c_nN + e_n + c_pP + e_p]$$
$$= [c_nc_pn_i^2(\exp(qV/kT) - 1)]/(c_pP_B)$$
$$= (N_B/\tau_B)(\exp(qV/kT) - 1)$$

$$R_N(1s) = [c_{ns}c_{ps}n_i^2(\exp(qV/kT) - 1)N_{IT}'(\text{bulk})]/[c_{ns}N_S+e_{ns}+c_{ps}P_S+e_{ps}]$$
$$= c_{ps}P_SN_{IT}'(\text{bulk})(\exp(qV/kT) - 1)$$
$$= (P_S/\tau_S)(\exp(qV/kT) - 1)$$

$$R_N(2b) = (n_i/\tau_{sc\text{-}ch})(\exp(qV/2kT) - 1)$$

$$I_Y = Z\int J_{NY}(x,y)dx \qquad \text{(P950.2B)}$$
$$= Z\int q\mu_n(dV/dy)Ndx$$
$$= Z(dV/dy)\mu_n\int qNdx$$
$$= Z\mu_n(dV/dy)Q_N(Q_{OT},Q_{IT},V_G)$$

Substitute (P950.2B) into (P950.2A):

$$(\partial/\partial y)\int J_{NY}Zdx = \partial I_Y/\partial y = \int qR_NZdx$$

$$\partial I_Y/\partial y = qZ\int\{(N_B/\tau_B)[\exp(qV/kT)-1] + (P_S/\tau_S)[\exp(qV/kT)-1] + (n_i/\tau_{sc\text{-}ch})[\exp(qV/2kT) - 1]\}dx$$

$$= qZ\{(N_BX_B/\tau_B) + P_Sc_{ps}\int N_{IT}'(\text{bulk})\delta(x)dx)[\exp(qV/kT) - 1] + (n_ix_{sc\text{-}ch}/\tau_{sc\text{-}ch})[\exp(qV/2kT) - 1]\}$$

Let $\bar{A} = (N_BX_B/\tau_B + P_Sc_{ps}\int N_{IT}'(\text{bulk})\delta(x)dx)$.

Note that $\int N_{IT}'(\text{bulk})\delta(x)dx = N_{IT}(V_S)\text{cm}^{-2}$.

Neglect recombination in the space–charge layer beneath the channel which is not expected to change with stress. The primary mechanism is N_{IT} generation at the SiO_2/Si interface and charging-discharging Q_{OT}.

$$\partial(I_Y/Z)/\partial y = q\bar{A}(\exp(qV/kT) - 1) \qquad \text{(P950.2A1)}$$

$$I_Y/Z = \mu_nQ_N(Q_{OT},Q_{IT},V_G)(dV/dy) \qquad \text{(P950.2B1)}$$

Substitute (P950.2B1) into (P950.2A1):

$$\partial(I_Y/Z)/\partial y = \mu_nQ_N(Q_{OT},Q_{IT},V_G)(\partial^2V/\partial y^2) = q\bar{A}(\exp(qV/kT) - 1)$$

$$(\partial^2V/\partial y^2) = (q\bar{A}/(\mu_nQ_N))(\exp(qV/kT) - 1)$$

Integrate by quadrature,

$$\partial^2V/\partial y^2 = (\partial V/\partial y)(\partial/\partial V)(\partial V/\partial y) = {}^1\!/_2(\partial/\partial V)(\partial V/\partial y)^2$$

$$\int d(\partial V/\partial y)^2 = 2\int(q\bar{A}/(\mu_nQ_N))(\exp(qV/kT) - 1)dV$$

$$(\partial V/\partial y)^2 = (2q\bar{A}/(\mu_nQ_N))[(kT/q)\exp(qV/kT) - (kT/q) - V]$$

$$\partial V/\partial y = \sqrt{[(2q\bar{A}/(\mu_nQ_N))(kT/q)[\exp(qV/kT)-(qV/kT)]]}$$

Now, we can apply the boundary condition to obtain I_{CH}.

$$I_{CH} = I_Y(0) = Z\mu_nQ_N(Q_{OT},Q_{IT},V_G)(\partial V/\partial y)|_{y=0}$$

$$= (Z\mu_nQ_N)\{(2q\bar{A}/\mu_nQ_N)(kT/q)[\exp(qV_{EB}/kT)-1-(qV_{EB}/kT)]\}^{1/2}$$

$$= Z\{(2q\bar{A}\mu_nQ_N)(kT/q)[\exp(qV_{EB}/kT)-1-(qV_{EB}/kT)]\}^{1/2}$$

where $\bar{A} = N_BX_B/\tau_B + P_Sc_{ps}N_{IT}$.

In order to calculate ΔI_{CH} and estimate TTF, we will need to make the following observations. When the BJT is stressed, N_{IT} becomes large.

$$\bar{A} \simeq P_Sc_{ps}N_{IT}$$

$$\Delta\bar{A} = P_Sc_{ps}\Delta N_{IT} = P_Sc_{ps}N_{IT\infty}[1 - \exp(-t/\tau)]$$

$$\Delta I_{CH} = I_{CH}(t_{stress}) - I_{CH}(t=0)$$

$$= + Z\sqrt{[\bar{A}(\text{stress})]}\sqrt{\{(2q\mu_nQ_N)(kT/q)[\exp(qV_{EB}/kT)-1-(qV_{EB}/kT)]\}} - Z\sqrt{[\bar{A}(t=0)]}\cdot\sqrt{\{(2q\mu_nQ_N)(kT/q)[\exp(qV_{EB}/kT)-1-(qV_{EB}/kT)]\}}$$

$$= + Z\{\sqrt{[\bar{A}(\text{stress})]} - \sqrt{[\bar{A}(t=0)]}\}\times$$

$\times\sqrt{\{(2q\mu_nQ_N)(kT/q)[\exp(qV_{EB}/kT)-1-(qV_{EB}/kT)]\}}$

We shall next make the assumption that at t=0, N_{IT0}=0, and no surface channel exists. Thus, $\mathbb{A}(t=0) = 0$.

$\Delta I_{CH} = Z\{[\mathbb{A}(stress)](2q\mu_nQ_N)(kT/q)[\exp(qV_{EB}/kT)-1-(qV_{EB}/kT)]\}^{1/2}$

$= Z\{[1-\exp(-t/\tau)](P_Sc_{ps}N_{IT\infty})(2q\mu_nQ_N)(kT/q)[\exp(qV_{EB}/kT)-1-(qV_{EB}/kT)]\}^{1/2}$

Note that the time dependence of $Q_N(t)$ is neglected to obtain an analytical solution. In actuality, $Q_N(t_{stress})$ due to positive oxide trap charging must be included.

$\Delta I_B(stress) = (qA_E\Delta S_EP_E)[\exp(qV_{EB}/kT)-1] + \Delta I_{CH}(1-\gamma_{E-ch}\alpha_{BF-ch})$

At the polysilicon/emitter interface, damage is small at moderate current during reverse–bias stress. Thus, $\Delta S_E \simeq 0$. Then,

$\Delta I_B(stress) \simeq \Delta I_{CH}(1-\gamma_{E-ch}\alpha_{BF-ch})$

$= Z\{[1-\exp(-t/\tau)](P_Sc_{ps}N_{IT\infty})(2q\mu_nQ_N)(kT/q)[\exp(qV_{EB}/kT)-1-qV_{EB}/kT)]\}^{1/2}$

$\times(1-\gamma_{E-ch}\alpha_{BF-ch}) = \sqrt{[1-\exp(-t/\tau)]}\mathbb{B}$

Let $\mathbb{B} = Z\{(P_Sc_{ps}N_{IT\infty})(2q\mu_nQ_N)(kT/q)[\exp(qV_{EB}/kT)-1-qV_{EB}/kT)]$
$\times(1-\gamma_{E-ch}\alpha_{BF-ch})\}^{1/2}$

We solve for the TTF using the failure criteria assumed for $\Delta h_{FE}/h_{FE}$.

$\Delta h_{FE}/h_{FE} = -\Delta I_B/I_B = -(\sqrt{[1-\exp(-t/\tau)]}\mathbb{B}/I_B)$

$1-\exp(-t/\tau) = [(\Delta h_{FE}/h_{FE})(I_B/\mathbb{B})]^2$

$\exp(-t/\tau) = 1-[(\Delta h_{FE}/h_{FE})(I_B/\mathbb{B})]^2$

Since t=TTF when $\Delta h_{FE}/h_{FE}$ = failure criterion = $\Delta h_{FE}(TTF)/h_{FE}$,

$TTF = -\tau\cdot\log_e\{1-[(\Delta h_{FE}/h_{FE})(I_B/\mathbb{B})]^2\}$.

Here, τ is given by the stress condition,

$\tau = q/[J_{G\text{-}stress}\sigma_n(E_0)]$

and the $\log_e\{\ldots\}$ term is determined at the operating point,

$\log_e\{1-[(\Delta h_{FE}(TTF)/h_{FE})(I_B(V_{EB},V_{CB})/\mathbb{B}(V_{EB},V_{CB}))]^2\}$.

Assume an operating point of V_{EB}=0.4V and V_{CB}=0.

$I_B/\mathbb{B} = A_E\{[(qN_BD_B/X_B)(X_B^2/(2D_B\tau_B)] + qP_ES_{E0}\}[\exp(qV_{EB}/kT)-1]$
$+ (qn_iX_{EB}/\tau_{EB})[\exp(qV_{EB}/2kT)-1]\}/Z\sqrt{\{(P_Sc_{ps}N_{IT\infty})(2q\mu_nQ_N)(kT/q)\times}$
$\times[\exp(qV_{EB}/kT)-1-qV_{EB}/kT)](1-\gamma_{E-ch}\alpha_{BF-ch})\}$

where $A_E = W_EZ$ so $A_E/Z = W_E$ and Z is normalized out. I_{CBO} is small at the operating point so neglected to first approximation. List the given parameters and compute the necessary parameters in the above equation.

$W_E = 1\mu m = 1\times10^{-4}$ cm

$N_B = n_i^2/P_{BB} = 10^{20}/10^{18} = 10^2$ cm^{-3}

$D_B = (kT/q)\mu_n$ (at base doping of $P_{BB}=10^{18}\ cm^{-3}$)

$= (0.0259V)(300\ cm^2/V\text{-}s)$ using Fig.313.5 FSSE $= 7.77\ cm^2/s$

$X_B = 0.1\ \mu m$

$\tau_B = 1/(c_{ne}N_{TT})$ (372.3)

Given $c_{ne} = 10^{-9}\ cm^3/s$ and assuming $N_{TT}=10^{12}\ cm^{-3}$,

$\tau_B = 10^{-3}\ s$

$P_E = n_i^2/N_{EE} = 10^{20}/10^{20} = 1\ cm^{-3}$.

$S_{E0} = 10^3\ cm/s$

$\tau_{EB} = \tau_{p0} + \tau_{n0} = 10^{-3}\ s + 10^{-3}\ s = 2\times10^{-3}\ s$

$X_{EB} = \sqrt{[2\varepsilon_s(V_{bi}-V_{EB})/(qN_M)]}$ (531.7)

The depletion approximation for X_{EB} is used, although at forward bias, an exact solution requires the electron and hole concentrations to be taken into account which may be done numerically.

$N_M = N_{AAE}N_{DDB}/(N_{AAE}+N_{DDB}) = (10^{20}10^{18})/(10^{20}+10^{18}) = 9.9\times10^{17}cm^{-3}$

$V_{biEB} = (kT/q)\log_e(N_{DDE}N_{AAB}/n_i^2)$

$= (0.0259V)\log_e(10^{20}10^{18}/10^{20}) = 1.073V$.

$X_{EB} = [(2\times11.8\times8.85\times10^{-14}F/cm)(1.073-0.4)\times$

$(1.602\times10^{-19}\times9.9\times10^{17}cm^{-3})^{-1}]^{1/2}$

$= 2.978\times10^{-6}\ cm = 0.030\ \mu m$

P_S = surface concentration of holes that are injected from the p–base into the n–channel emitter layer. In a P/N junction, this can be evaluated by multiplying the minority carrier concentration at the edge of the SCL with the Boltzmann factor.

$P(x) = n_i\exp\{(q/kT)(V_P(x)-V_I(x))\}$.

In the channel, the hole concentration must be determined at a point, y.

At y=0, at the emitter edge,

$P(x=x_n) = P_E\exp(qV_{EB}/kT) = (1.0cm^{-3})\exp(0.4V/0.0259V) = 5.1\times10^6cm^{-3}$.

Assume $P_S \simeq P(x=x_n, y=0) = 5.1\times10^6cm^{-3}$. A more accurate 2–dimensional analysis can be made for $P_S(y)$ by taking into account the surface potential variation along the channel. At $y = \infty$,

$P_S \rightarrow P_{BB} = 10^{18}\ cm^{-3}$.

$Q_N = C_o[V_G - V_{GT} - V(y)]$ (642.2)

The surface will become inverted after stress when positive charge is trapped in the oxide. We shall neglect the time dependence of Q_N due to the trapping of oxide charge. An approximate solution is obtained by calculating Q_N when $N_{OT\infty} = 10^{12}\ cm^{-2}$ oxide traps are charged positively. The time dependence in this solution is determined by the N_{IT} generation rate.

V_{Fbase} = $(kT/q)\log_e(P_{BB}/n_i) = (0.0259V)\log_e(10^{18}/10^{10}) = 0.477V$

V_{GTS} = **threshold gate voltage relative to source**

$= V_{FB} + 2V_F + \surd[2\varepsilon_s qN_{AA}(V_S-V_X+2V_F)]/C_o$ (643.3)

where

$C_o = \varepsilon_o/x_0 = 3.45\times10^{-7}$ F/cm^2.

$V_{FB} = \phi_{GX} - Q_{OT}/C_o \rightarrow \phi_{GX} - qN_{OT\infty}/C_o$

$= \phi_{GX} - (1.602\times10^{-19}C)(10^{12}\ cm^{-2})/(3.45\times10^{-7}\ F/cm^2)$

$= \phi_{GX} - 0.4639V$

For the n+ polysilicon gate ($N_{GG}=10^{20}cm^{-3}$, N_{BB}=p–substrate=$10^{18}cm^{-3}$),

$\phi_{GX} = -(kT/q)\log_e(N_{GG}P_{XX}/n_i^2) = -(kT/q)\log_e(N_{GG}P_{BB}/n_i^2)$

$= -(0.259V)\log_e(10^{20}10^{18}/10^{20}) = -1.073V.$

$V_{FB} = -1.073 - 0.4639 = -1.537V$

$V_{GTS} = V_{FB} + 2V_F + \surd[2\varepsilon_s qN_{AA}(V_S-V_X+2V_F)]/C_o$

$= -1.537 + 2(0.477)$

$+ \{2\times11.8\times8.854\times10^{-14}(F/cm)\times1.602\times10^{-19}C\times10^{18}cm^{-3}$

$[0-0 + 2(0.477V)]\}^{1/2}/3.45\times10^{-7}\ F/cm^2 = 1.054V$

This solution is too large and the surface is not inverted at $V_G=V_{EB}=0.4V$ which is the assumed operating point where $\Delta h_{FE}/h_{FE}$ is measured. Thus, we increase $N_{OT\infty}$ from 1×10^{12} cm^{-2} to 4×10^{12} cm^{-2}.

$V_{FB} = \phi_{GX} - Q_{OT}/C_o \rightarrow \phi_{GX} - qN_{OT\infty}/C_o$

$= -1.073-(1.602\times10^{-19}C)(4\times10^{12}cm^{-2})/(3.45\times10^{-7}F/cm^2)$

$= -2.9287V$

$V_{GTS} = -2.9287+2(0.477)+1.6365 = -0.338V$ (surface is inverted)

At y=0, $Q_N = C_o[V_G-V(y=0)-V_{GTS}] = C_o[0.4-0.4-(-0.338)]=1.16\times10^{-7}C/cm^2$

At y=∞, $Q_N = C_o[V_G-V(y=\infty)-V_{GTS}] = C_o[0.4-0.0-(-0.338)]=2.55\times10^{-7}C/cm^2$

Emitter injection efficiency calculated from Gummel numbers,

$\gamma_{E-ch} = j_B/(j_E + j_{EB} + j_B) = 1/(j_E/j_B + j_{EB}/j_B + 1)$ (737.13)

$j_E/j_B = (D_E/D_B)(N_{BB}X_B)/(N_{EE}/X_E) = (D_E/D_B)(G_B/G_E)$ (737.14)

In the surface channel,

$D_B = \mu_n(\text{at } P_{BB}=10^{18}cm^{-3})(kT/q) = (300cm^2/V\text{–}s)(0.0259V) = 7.77cm^2/s$

$D_E = \mu_p(\text{at } P_{BB}=10^{18}cm^{-3})(kT/q) = (100cm^2/V\text{–}s)(0.0259V) = 2.59cm^2/s$

since the emitter is the inversion channel in the p–base layer.

$j_E/j_B = (2.59/7.77)(10^{18}\times0.1\times10^{-4})/(Q_N/q)$

$= (2.59/7.77)(10^{18}cm^{-3}\times0.1\times10^{-4}cm)(1.60\times10^{-19}C)/(1.16\times10^{-7}C/cm^2)$

$= 4.60$

$j_{EB}/j_B = [(qn_iX_{EB}/\tau_{EB})/(qD_BP_B/L_B)]$

$\times[(\exp(qV_{EB}/2kT)-1)/((\exp(qV_{EB}/kT)-1)\text{ctnh}(X_B/L_B)]$

$\simeq ((n_iX_{EB}X_B)/(D_BN_B\tau_{EB}))(1/\exp(qV_{EB}/2kT))$

$= [(10^{10}\text{cm}^{-3})(2.9787\text{x}10^{-6}\text{cm})(0.1\text{x}10^{-4}\text{cm})]/$

$[(7.77\text{cm}^2/\text{s})(10^2\text{cm}^{-3})(2\text{x}10^{-3}\text{s})\exp(0.4/2\text{x}0.0259)]$

$= 8.49\text{x}10^{-5}$

$\gamma_{E\text{-}ch} = 1/(j_E/j_B + j_{EB}/j_B + 1) = 1/(4.60+8.49\text{x}10^{-5}+1) = 0.179$

Base transport factor,

$\alpha_{BF\text{-}ch} = \text{sech}(X_B/L_B)$ (737.2)

$\simeq 1 - X_B^2/(2D_B\tau_B)$ (737.2B)

$= 1 - (0.1\text{x}10^{-4}\text{cm})^2/[2\text{x}7.77(\text{cm}^2/\text{s})\text{x}10^{-3}\text{s}] = 1.00000$

We can now obtain an estimate for

$I_B/\mathbf{B} = A_E\{[(qN_BD_B/X_B)(X_B^2/(2D_B\tau_B)] + qP_ES_{E0}][\exp(qV_{EB}/kT)-1]$

$+ (qn_iX_{EB}/\tau_{EB})[\exp(qV_{EB}/2kT) - 1]\}\times(1/Z)$

$\times \{(P_Sc_{ps}N_{IT\infty})(2q\mu_nQ_N)(kT/q)[\exp(qV_{EB}/kT)-1-(qV_{EB}/kT)](1 - \gamma_{E\text{-}ch}\alpha_{BF\text{-}ch}\}^{1/2}$

Calculate each part separately:

$(qN_BD_B/X_B) = (1.602\text{x}10^{-19}\text{C}\text{x}10^2\text{cm}^{-3}\text{x}7.77\text{cm}^2/\text{s})/(0.1\text{x}10^{-4}\text{cm})$

$= 1.2447\text{x}10^{-11}\ \text{C/cm}^2\text{–s}$

$X_B^2/(2D_B\tau_B) = (0.1\text{x}10^{-4}\text{cm})^2/[2\text{x}7.77(\text{cm}^2/\text{s})\text{x}10^{-3}\text{s}] = 6.435\text{x}10^{-9}$

$(qP_ES_{E0}) = (1.602\text{x}10^{-19})(1\text{cm}^{-3})(10^3\ \text{cm/s}) = 1.602\text{x}10^{-16}\ \text{C/cm}^2\text{–s}$

$(qn_iX_{EB}/\tau_{EB}) = (1.602\text{x}10^{-19}\text{C})(10^{10}\ \text{cm}^{-3})(2.978\text{x}10^{-6}\text{cm})/(2\text{x}10^{-3}\text{s})$

$= 2.385\text{x}10^{-12}\ \text{C/cm}^2\text{–s}$

$(P_Sc_{ps}N_{IT\infty})(2q\mu_nQ_N)(kT/q) = (5.1\text{x}10^6\text{cm}^{-3})(10^{-9}\text{cm}^3/\text{s})\times$

$\times(10^{12}\text{cm}^{-2})(2\text{x}1.602\text{x}10^{-19}\text{C}\text{x}300(\text{cm}^2/\text{V-s})\text{x}1.16\text{x}10^{-7}(\text{C/cm}^2)(0.0259\text{V})$

$= 1.473\text{x}10^{-15}\ \text{C/cm}^2\text{–s}^2$

$\exp(qV_{EB}/kT) = \exp(0.4/0.0259) = 5.096\text{x}10^6$

$\exp(qV_{EB}/2kT) = \exp(0.4/2\text{x}0.0259) = 2.257\text{x}10^3$

$(1 - \gamma_{E\text{-}ch}\alpha_{BF\text{-}c}) = 1 - 0.179\text{x}1 = 0.821$

$I_B/\mathbf{B} = (1\text{x}10^{-4}\text{cm})\{[(1.2447\text{x}10^{-11}\ \text{C/cm}^2\text{–s})(6.435\text{x}10^{-9})$

$+ (1.602\text{x}10^{-16}\text{C/cm}^2\text{–s})](5.096\text{x}10^6)$

$+ (2.387\text{x}10^{-12}\text{C/cm}^2\text{–s})(2.257\text{x}10^3)\}/$

$[(1.559\text{x}10^{-10}\text{C}^2/\text{cm}^2\text{–s}^2)(5.096\text{x}10^6\text{–}15.4)]^{1/2}(0.821)$

$= 8.721\text{x}10^{-9}$

This gives a very short lifetime. Increasing S_{E0} to 10^9cm/s gives $I_B/\mathbf{B} = 1.147\text{x}10^{-3}$.
Next, we find τ for the stress conditions.

(a) V_{EB} = 5V reverse bias

Calculate τ from $J_{-Gstress}$ due to electron tunneling from the substrate near the end of the channel.

As calculated above, $\phi_{GX} = -1.073V$ and $V_F = 0.477V$.

$V_G = V_{EB} = 5V$, Need to obtain V_O by iteration.

$V_{AA} = 0.708V$

$V_S^0 = 2V_F = 0.954V$

$\Delta_I^0 = 36.83397$ where $U_\beta = V_S^0/(kT/q) = 36.83397 = 2U_F$

$V_S^1 = 2V_F + (kT/q)\log_e\{[(V_G-V_{FB}-V_S^0)^2/(4(kT/q)V_{AA}\Delta_I^0)]\} = 0.954 +$
$+ (0.0259)\log_e\{[5-(-1.073)-0.954]^2/(4x0.0259x0.708x36.8)\} =$
$1.015V$

$Error^1 = V_S^1 - V_S^0 = 1.015 - 0.954 = 0.061V$

$\Delta_I^1 = 1.0000$

$V_S^2 = 0.954 + (0.0259)\log_e\{[5-(-1.073)-1.015]^2/(4x0.0259x0.708x1)\}$
$= 1.123\ V$

$Error^2 = V_S^2 - V_S^1 = 1.123 - 1.015 = 0.108\ V$

$\Delta_I^2 = 1.0000$

$V_S^3 = 0.954 + (0.0259)\log_e\{[5-(-1.073)-1.123]^2/(4x0.0259x0.708x1)\}$
$= 1.105\ V$

$Error^3 = V_S^3 - V_S^2 = 1.105 - 1.123 = -0.018V$

$\Delta_I^3 = 1.0000$

$V_S^4 = 0.954 + (0.0259)\log_e\{[5-(-1.073)-1.105]^2/(4x0.0259x0.708x1)\}$
$= 1.105\ V$

$Error^4 = V_S^4 - V_S^3 = 1.105 - 1.105 = 0.000$ Converged.

$V_O = V_G - V_{FB} - V_S = 5 - (-1.073) - 1.105 = 4.97V$

$E_O = V_O/x_O = 4.97\ MV/cm$

$J_G = 2.6x10^6 E_O^2 \exp(-238.5/E_O)$
$= 9.11x10^{-14}\ A/cm^2$

$\tau = q/(J_{Gstress}\sigma(E_O))$
$= 1.602x10^{-19}C/(9.11x10^{-14}A/cm^2 x10^{-16}cm^2)$
$= 1.758x10^{10}\ s$

$TTF = -\ \tau\log_e\{1 - [(\Delta h_{FE}/h_{FE})(I_B/B)]^2\}$
$= -\ (1.758x10^{10}s)\log_e\{1 - [(0.2)(1.147x10^{-3})]^2\}$
$= 1x10^3\ s$

(b) V_{EB} = 3.3V reverse bias

Similar iteration gives

V_S = 1.105V
V_0 = 3.27 V
E_0 = 3.37 MV/cm
J_G = 5.6×10^{-25} A/cm^2
τ = $(1.602\times10^{-19}C)/(5.6\times10^{-25}A/cm^2\times3.5\times10^{-16}cm^2) = 8.1\times10^{20}s$
TTF = $-(8.1\times10^{20}s)\log_e\{1 - [(0.2)(1.147\times10^{-3})]^2\}$
= 4×10^{13} s
= 1×10^{6} years (Will not fail.)

(c) V_{EB} = 2.5V reverse bias

Similar iteration gives

V_S = 1.105V
V_0 = 2.47 V
E_0 = 2.47 MV/cm
J_G = 1.7×10^{-35} A/cm^2
τ = $(1.602\times10^{-19}C)/(1.7\times10^{-35}A/cm^2\times7.9\times10^{-16}cm^2) = 1.2\times10^{31}s$
TTF = $-(1.2\times10^{31}s)\log_e\{1 - [(0.2)(1.147\times10^{-3})]^2\}$
= 6×10^{23} s = 2×10^{16} years (Will not fail.)

P950.3 Calculate the time–to–failure (TTF) due to acceptor hydrogenation of a MOS transistor under operating conditions using the physics of acceptor hydrogenation and the MOS transistor linear and subthreshold characteristics. (Oxide and interface trap charging is not included in this problem which were considered already in P950.1.) The following parameters are given.

x_{OX} = 100 A
μ_0 = 600 cm^2/V–s
N_{AA} = 10^{17} cm^{-3}
A(0) = N_{AA}
t_D = 0 (Assume no initial delay.)
k_H = 10×10^{-10} s^{-1} (at the operating voltage of 5V)
e_H = 1×10^{-10} s^{-1}
$c_H H_G$ = 9×10^{-10} s^{-1}

Calculate the TTF using the following two criteria (a) and (b):

(a) ΔV_T = 20 mV at an operating gate voltage of 5V.
(b) $\Delta I_D/I_D$ = 2% at an operating gate voltage of 5V.

The formulae for the acceptor hydrogenation and dehydrogenation were derived as a function of time. Write the acceptor hydrogenation equation as a function of current density injected into the gate oxide.

Solutions:

P950.3(a) $\Delta V_T = 20mV$

The hydrogenation and annealing kinetic model of group III acceptors was analyzed by Sah and two graduate students [930.2]. The reaction equations of the acceptor deactivation are given by:

$X{\bullet}H + e^{**}$	$\rightarrow X + H{\bullet} + e^{*}$	**Hydrogen bond breaking.**
$H{\bullet}(gate)$	$\rightarrow H{\bullet}(silicon)$	**Hydrogen migration through oxide.**
$h + Si{-}A{\equiv}S_{i3} + H{\bullet}$	$\rightarrow Si{\bullet}H{\bullet}A{\equiv}Si_3$	**Hydrogenation of acceptor.**
$H{\bullet} + H{\bullet}$	$\rightarrow H_2$	**Hydrogen recombination–generation.**
$Y + nH{\bullet}$	$\rightarrow Y{\bullet}H_n$	**Hydrogen trapping.**

The differential equations for the reaction are given by Eqs.(6)–(8b) in [930.2].

$$dH/dt|_{gate} = e_X(N_{XX} - X) - c_X XH$$
$$dH/dt|_{oxide} = D_H \nabla^2 H$$
$$dH/dt|_{silicon\text{-}1} = e_H(N_{AA} - A) - c_H HA = dA/dt$$
$$dH/dt|_{silicon\text{-}2} = g_H H_2 - r_H H^2$$
$$dH/dt|_{silicon\text{-}3} = e_Y(N_{YY} - Y) - c_Y YH^n$$

The long time solution for the electrically active acceptor concentration, A(t), in the substrate is given by Eq.(23)[930.2].

$$A(t) = A(0) + [A(\infty)-A(0)]\{1 - \exp[-(e_H+c_H H_G)(t-t_D)]\}$$

where $A(\infty)$ is the asymptotic electrically active acceptor concentration,

$$A(\infty) = (e_H/k_H)N_{AA}$$

A(0) is the total acceptor concentration, $A(0)=N_{AA}$, e_H is the hydrogen emission rate, and k_H is the first order hydrogenation reaction rate given by

$$k_H = e_H + c_H H_G$$

The threshold voltage of the MOST is given by

$$V_{GTS} = \text{threshold gate voltage relative to source}$$
$$= V_{FB} + 2V_F + \sqrt{[2\varepsilon_s q N_{AA}(V_S - V_X + 2V_F)]}/C_o \qquad (643.3)$$
$$V_{FB} = \phi_{GX} - (Q_{OT} + Q_{IT})/C_o = \phi_{GX}.$$

In this problem, the oxide and interface trap charge densities are assumed to be zero. Furthermore, let $V_S = V_X = 0V$.

The change in threshold voltage due to acceptor hydrogenation is given by

$$
\begin{aligned}
\Delta V_{GTS}(t) &= + \ \phi_{GX}(t) + 2V_F(t) + \surd[2\varepsilon_S qA(t)2V_F(t)]/C_o \\
&\quad - [\phi_{GX}(0) + 2V_F(0)] - \surd[2\varepsilon_S qA(0)2V_F(0)]/C_o \\
&= - (kT/q)\log_e[N_{GG}A(t)/n_i{}^2] + 2(kT/q)\log_e[A(t)/n_i] \\
&\quad + \surd[4\varepsilon_S qA(t)(kT/q)\log_e(A(t)/n_i)]/C_o \\
&\quad - [-(kT/q)\log_e(N_{GG}N_{AA}/n_i{}^2) \\
&\quad + 2(kT/q)\log_e(N_{AA}/n_i) + \surd[4\varepsilon_S qN_{AA}(kT/q)\log_e(N_{AA}/n_i)]/C_o \\
&= + (kT/q)\log_e(N_{AA}/A) + 2(kT/q)\log_e(A/N_{AA}) \\
&\quad + (2/C_o)\surd[\varepsilon_S q(kT/q)]\{\surd A\log_e(A/n_i) \\
&\qquad\qquad - \surd[N_{AA}\log_e(N_{AA}/n_i)]\} \\
&= + (kT/q)\log_e(A/N_{AA}) \\
&\quad + \{\surd[A\log_e(A/n_i)] - \surd[N_{AA}\log_e(N_{AA}/n_i)]\}\times \\
&\quad \times \{(2/C_o)\surd[\varepsilon_S q(kT/q)]\}
\end{aligned}
$$

The following equation may be numerically solved to obtain the TTF:

$$
\begin{aligned}
\Delta V_{GTS}(TTF) &= (kT/q)\log_e(A(TTF)/N_{AA}) + \{\surd[A(TTF)\log_e(A(TTF)/n_i)] \\
&\quad - \surd[N_{AA}\log_e(N_{AA}/n_i)]\}\{(2/C_o)\surd[\varepsilon_S q(kT/q)]\}
\end{aligned}
$$

using the threshold voltage failure criteria, $|\Delta V_{GTS}(TTF)| = |-20\ mV|$. The sign of $\Delta V_{GTS}(TTF)$ is negative for acceptor hydrogenation.

An approximate analytical solution may be obtained by noting that $\log_e(A(TTF)/N_{AA}) \simeq 0$ and $\log_e(A(TTF)/n_i) \simeq \log_e(N_{AA}/n_i)$. Then,

$$
\begin{aligned}
\Delta V_{GTS}(TTF) &\simeq \{(2/C_o)\surd[\varepsilon_S q(kT/q)\log_e(N_{AA}/n_i]\}\{\surd[A(TTF)] - \surd[N_{AA}]\} \\
A(TTF) &= \{\Delta V_{GTS}(TTF)/[(2/C_o)\surd[\varepsilon_S q(kT/q)\log_e(N_{AA}/n_i]] + \surd[N_{AA}]\}^2 \\
A(0) + [A(\infty) - A(0)]\{1 - \exp(-k_H TTF)\} &= \\
&\{\Delta V_{GTS}(TTF)/[(2/C_o)\surd[\varepsilon_S q(kT/q)\log_e(N_{AA}/n_i]] + \surd[N_{AA}]\}^2 \\
1-\exp[-k_H TTF] &= \{\{\Delta V_{GTS}(TTF)/[(2/C_o)\surd\varepsilon_S q(kT/q)\log_e(N_{AA}/n_i)] \\
&\quad + \surd N_{AA}\}^2 - N_{AA}/[(e_H/k_H)N_{AA} - N_{AA}] \\
-k_H TTF &= \log_e\{1-(1/N_{AA})\Delta V_{GTS}(TTF)/[(2/C_o)\surd[\varepsilon_S q(kT/q)\log_e(N_{AA}/n_i]] \\
&\quad + \surd N_{AA}\}^2 - 1\}/[(e_H/k_H) - 1]\}
\end{aligned}
$$

Compute part of the above equation separately:

$$
\begin{aligned}
&(2/C_o)\surd[\varepsilon_S q(kT/q)\log_e(N_{AA}/n_i]] = \\
&\quad (2/3.453\times10^{-7}\ F/cm^2)[11.8\times8.854\times10^{-14}(F/cm)\times1.602\times10^{-19}C\times0.41746V]^{1/2} \\
&\quad = 1.531\times10^{-9}\ Vcm^{3/2}. \\
&TTF = (-1/10^{-9} \\
&\qquad + \surd(10^{17}cm^{-3})]^2-1\}/[(1/10)- 1]\} \\
&\quad = (-1/10^{-9}\ s^{-1})\log_e\{1 - 0.0899\} = 9.4\times10^7\ s = 3.0\ years
\end{aligned}
$$

P950.3(b) $\Delta I_{D\text{-}sat}/I_{D\text{-}sat} = 2\%$

$$I_{D\text{-}sat} = (Z/L)(\mu_n C_o/2)(V_G - V_{GT})^2 \qquad (643.6)$$

$$\Delta I_{D\text{-}sat} = (\partial I_{D\text{-}sat}/\partial\mu_n)d\mu_n + (\partial I_{D\text{-}sat}/\partial V_T)dV_T$$

$$= (Z/L)(C_o/2)(V_G - V_{GT})^2 d\mu_n + 2(Z/L)(\mu_n C/2)(V_G - V_{GT})(-1)dV_{GT}$$

$$\Delta I_{D\text{-}sat}/I_{D\text{-}sat} = [(Z/L)(C_o/2)(V_G - V_{GT})^2 \Delta\mu_n$$

$$+ 2(Z/L)(\mu_n C/2)(V_G - V_{GT})(-1)\Delta V_{GT}]\times$$

$$\times [(Z/L)(\mu_n C_o/2)(V_G - V_{GT})^2]^{-1}$$

$$= \Delta\mu_n/\mu_n - 2\Delta V_{GT}/(V_G - V_{GT})$$

where

$$1/\mu_n = 1/\mu_{nI} + 1/\mu_{nL}$$

$$= 1/\mu_{nI} + 1/\mu_{nA} + 1/\mu_{nO} \quad \text{(n=electron mobility)} \qquad (313.12)$$

where μ_{nI} is the electron mobility due to scattering by ionized impurity, μ_{nA} due to lattice scattering by acoustic phonons, and μ_{nO} due to lattice scattering by optical phonons.

$$\mu_n = \mu_{nI}\mu_{nL}/(\mu_{nL} + \mu_{nI}) \text{ where}$$

$$1/\mu_{nL} = 1/\mu_{nA} + 1/\mu_{nO}$$

The ionized impurity scattering-limited mobility, μ_{nI}, will change as a function of time as the number of electrically active ionized acceptor impurities, A(t), decreases due to hydrogenation.

$$\mu_{nI} = \mu_\infty + \mu_{I0}(N_0/N_I)^\alpha(T/300)^\beta \qquad \text{(Table 313.1 FSSE, p.250)}$$

$$\mu_{nI}(t) = \mu_\infty + \mu_{I0}(N_0/A(t))^\alpha(T/300)^\beta$$

$$\Delta\mu_n = \mu_n(t) - \mu_n(0)$$

$$= \{\mu_{nI}(t)\mu_{nL}/(\mu_{nL} + \mu_{nI}(t)) - \mu_{nI}(0)\mu_{nL}/(\mu_{nL} + \mu_{nI}(0))\}$$

$$\Delta\mu_n/\mu_n = (\mu_n(t) - \mu_n(0))/\mu_n(0)$$

$$= \mu_n(t)/\mu_n(0) - 1$$

Now consider the second term in the expression for $\Delta I_{D\text{-}sat}/I_{D\text{-}sat}$.

$$\Delta I_{D\text{-}sat}/I_{D\text{-}sat} = \Delta\mu_n/\mu_n - 2\Delta V_{GT}/(V_G-V_{GT})$$

$$V_{GTS} = \text{threshold gate voltage relative to source}$$

$$= V_{FB} + 2V_F + \surd[2\varepsilon_S qN_{AA}(V_S-V_X+2V_F)]/C_o \qquad (643.3)$$

$$V_{FB} = \phi_{GX} - (Q_{OT} + Q_{IT})/C^o \equiv \phi_{GX} - (0+0)/C_o$$

$$\phi_{GX} = -(kT/q)\log_e(N_{GG}P_{XX}/n_i^2). \quad \text{Let } N_{GG} = 10^{20}\text{cm}^{-3},$$

$$\phi_{GX} = -(0.0259V)\log_e(10^{20}10^{17}/10^{20}) = -1.014V$$

$$V_F = (kT/q)\log_e(N_{AA}/n_i) = (0.0259V)\log_e(10^{17}/10^{10}) = 0.4174V$$

$$\surd[2\varepsilon_S qN_{AA}(V_S-V_X+2V_F)]/C_o = (2/3.453\text{x}10^{-7}F/\text{cm}^2)\times$$

$$[11.8\text{x}8.854\text{x}10^{-14}\text{x}1.602\text{x}10^{-19}C\text{x}10^{17}\text{cm}^{-3}\text{x}0.4174V]^{1/2} = 0.4841V$$

$$V_{GTS} = -1.014 + 2(0.4174) + 0.4841 = 0.305V$$

The TTF is obtained by numerically solving

$$\Delta I_{D\text{-}sat}(TTF)/I_{D\text{-}sat} = \Delta\mu_n(TTF)/\mu_n - 2\Delta V_{GT}(TTF)/(V_G-V_{GT})$$

where

$$\Delta\mu_n(TTF)/\mu_n =$$

$$\{\mu_\infty+\mu_{I0}(N_0/A(TTF))^\alpha(T/300)^\beta)\mu_{nL}/(\mu_{nL}+(\mu_\infty+\mu_{I0}(N_0/A(TTF))^\alpha(T/300)^\beta))\}/$$

$$\{(\mu_\infty+\mu_{I0}(N_0/N_{AA})^\alpha(T/300)^\beta)\mu_{nL}/(\mu_{nL}+(\mu_\infty+\mu_{I0}(N_0/N_{AA})^\alpha(T/300)^\beta))\} - 1$$

and

$$\Delta V_{GTS}(TTF) = (kT/q)\log_e(A(TTF)/N_{AA})$$

$$+ \{\surd[A(TTF)\log_e(A(TTF)/n_i)] - \surd[N_{AA}\log_e(N_{AA}/n_i)]\}\times$$

$$\times \{(2/C_o)\surd[\varepsilon_S q(kT/q)]\}.$$

To compute the TTF, use the failure criteria given by,

$$\Delta I_{D\text{-}sat}(TTF)/I_{D\text{-}sat} = 0.02 \text{ or } 2\%.$$

This gives

$$TTF = 3.889\text{x}10^7 s = 1.2 \text{ years}$$

Use the assumed value of hydrogenation reaction rate, k_H, and emission rate, e_H. At t=TTF, the electrically active acceptor concentration has decreased from 10^{17} cm^{-3} to 9.657x10^{16} cm^{-3}, the mobility has increased from 660 cm^2/V-s to 670 cm^2/V-s, and the gate threshold voltage has decreased from 0.305 to 0.294V.

P950.4 Calculate the time-to-failure due to electromigration estimated using the Black equation. Data is obtained for a Ti:W/Al-Cu(0.5%)-Si(1%) metallization layer with resistivity, ρ=5$\mu\Omega$-cm. At 1.0×10^6A/cm^2, MTF=1600hr at 150°C, E_A=0.6 eV, and n=2. The failure criteria is usually defined as $\Delta R/R \geq 10\%$; however a more important failure criteria for circuit applications is the change in propagation delay.

At the normal operating condition, I=2mA flows in the interconnect line of dimensions, L_W=10μm, W_W=0.6μm, t_{metal}=0.5μm, and t_i=0.5μm at 50°C.

P950.4(a) Calculate the propagation delay for three cases of inter-metal dielectric: **(i)** vacuum (limiting case), **(ii)** SiO_2 (ε_o=3.9), **(iii)** Si_3N_4 (ε_{Si3N4}=7.5).

(i) Vacuum as dielectric layer (ε=1)

$$\tau_{RC}/L^2 = (\varepsilon_o/t_i)(\rho/t_m)$$
$$= (1.0\times8.854\times10^{-14}\ \mathrm{F/cm}/0.5\times10^{-4}\mathrm{cm})(5\times10^{-6}\Omega\text{-cm}/0.5\times10^{-4}\mathrm{cm})$$
$$= (1.7708\times10^{-9}\mathrm{F/cm^2})(0.1\Omega/\mathrm{square})$$
$$= 1.7708\times10^{-10}\mathrm{s/cm^2}$$

For L_W=10μm length of interconnect,

$$\tau_{RC}(\mathrm{vacuum}) = (1.7708\times10^{-10}\ \mathrm{s/cm^2})(10\times10^{-4}\mathrm{cm})^2 = 1.7708\times10^{-16}\mathrm{s}.$$

(ii) Silicon dioxide dielectric. SiO_2 ($\varepsilon^o{}_x$=3.9)

$$\tau_{RC}/L^2 = (\varepsilon_{ox}/X_i)(\rho/X_m)$$
$$= (3.9\times8.854\times10^{-14}\mathrm{F/cm}/0.5\times10^{-4}\mathrm{cm})(5\times10^{-6}\Omega\text{-cm}/0.5\times10^{-4}\mathrm{cm})$$
$$= (6.906\times10^{-9}\mathrm{F/cm^2})(0.1\Omega/\mathrm{square})$$
$$= 6.906\times10^{-10}\mathrm{s/cm^2}$$

For L_W=10μm length of interconnect,

$$\tau_{RC}(SiO_2) = (6.906\times10^{-10}\ \mathrm{s/cm^2})(10\times10^{-4}\mathrm{cm})^2 = 6.906\times10^{-16}\mathrm{s}.$$

(iii) Silicon nitride dielectric. Si_3N_4 (ε_{Si3N4}=7.5)

$$\tau_{RC}/L^2 = (\varepsilon_{Si3N4}/X_i)(\rho/X_m)$$
$$= (7.5\times8.854\times10^{-14}\mathrm{F/cm}/0.5\times10^{-4}\mathrm{cm})(5\times10^{-6}\Omega\text{-cm}/0.5\times10^{-4}\mathrm{cm})$$
$$= (1.328\times10^{-8}\mathrm{F/cm^2})(0.1\Omega/\mathrm{square})$$
$$= 1.328\times10^{-9}\mathrm{s/cm^2}$$

For L_W=10μm length of interconnect,

$$\tau_{RC}(Si_3N_4) = (1.328\times10^{-9}\ \mathrm{s/cm^2})(10\times10^{-4}\mathrm{cm})^2 = 1.328\times10^{-15}\mathrm{s}.$$

P950.4(b) Calculate the TTF with a void present. If a void of dimensions 0.5μm×0.5μm ($L_V \times W_V$) and depth 0.5μm is opened in the center of the interconnect initially due to a dust particle during photolithography, estimate the TTF due to electromigration.

The highest current density occurs where the void is located which determines the TTF.

$$A_{metal\_min} = X_m \cdot (W - 0.5\mu m)$$
$$= (0.5\times10^{-4}\ cm)(0.6\times10^{-4}\ cm - 0.5\times10^{-4}\ cm)$$
$$= 5\times10^{-10} cm^2$$

I = 2mA given at the operating condition

$$J_{operating\_condition\_max} = I/A_{metal\_min}$$
$$= (2\times10^{-3}A)/(5\times10^{-10}\ cm^2)$$
$$= 4\times10^{6} A/cm^2$$

The TTF is estimated using the empirical Black equation

$$MTTF = AJ^{-n}\exp(E_A/kT)$$

From experimental data given, MTTF=1600hr at $J_{stress}=1\times10^6 A/cm^2$, T=150°C, n=2, E_A=0.6 eV. Solve for the pre-exponential factor, A:

$$A = MTTF/[J^{-n}\exp(E_A/kT)] \text{ where } k = 8.616\times10^{-5}\ eV/K.$$
$$= (1600\ hr)(1\times10^{6}\ A/cm^2)^2\times$$
$$\times \exp\{-0.6eV/[(8.616\times10^{-5}eV/K)(150+273.15K)]\}$$
$$= 1.15\times10^{8}\ hr\cdot A^2/cm^4$$

Now, calculate the TTF at the operating condition of I=2 mA and T=50°C, taking into account the increased current density caused by the void.

$$J_{operating\_condition\_max} = I/A_{metal\_min} = 4\times10^{6}\ A/cm^2$$

$$MTTF = AJ^{-n}\exp(E_A/kT)$$
$$= (1.15\times10^{8}\ hr\cdot A^2/cm^4)(4\times10^{6}\ A/cm^2)^{-2}\times$$
$$\times \exp\{0.6eV[(8.616\times10^{-5}eV/K)(50+273.15K)]^{-1}\}$$
$$= 1.62\times10^{4}\ hr = 1.85\ years$$

P950.4(c) Calculate the propagation delay of the interconnect in (b).

Assume SiO_2 dielectric, then

$$t_{RC} = R\times(C_{ox}A_{ox}) \text{ where}$$
$$R = R_1 + R_{void}$$
$$= \rho(L-X_{void})/(W\times X_m) + \rho\times X_{void}/[(W-X_{void})X_m]$$
$$= (5\times10^{-6})10^4(10-0.5)/(0.6\times0.5)$$
$$+ (5\times10^{-6})10^4(0.5)\times[(0.6-0.5)(0.5)]^{-1}$$
$$= 1.583\Omega + 0.500\Omega = 2.083\ \Omega$$
$$t_{RC} = R\times(C_{ox}A_{ox})$$
$$= R\times C_{ox}\times(WL - x_{void}^2)$$

where

$$C_{ox} = \varepsilon_{ox}/X_{ox}$$
$$= (3.9\times8.854\times10^{-14}F/cm)/(0.5\times10^{-4}cm) = 6.906\times10^{-9}\ F/cm^2$$
$$\tau_{RC} = (2.083)(6.906\times10^{-9})10^{-8}(0.6\times10-0.5^2)$$
$$= 8.27\times10^{-16}\ s$$

Compare this with the t_{RC} calculated in part (a) with no void.

$$100\times(t_{RC\_void} - t_{RC\_no\_void})/t_{RC\_no\_void}$$
$$= 100\times(8.27\times10^{-16}s - 6.91\times10^{-16}s)/(6.91\times10^{-16}s)$$
$$= 20\% \text{ increase}$$

The propagation delay is increased 20% by void formation which reduces the operation lifetime or time-to-failure due to electromigration at high current densities. The void-induced higher current density increases the rate of formation of the void due to both higher current density and Joule heating which further increase the current density. This is a regenerative feedback loop which leads eventually to catastrophic failure due to open-circuit of the interconnect line and disconnection of the adjacent transistors. The localized melting of the aluminum around the void may also alloy into adjacent SiO_2 thereby causing short-circuit to adjacent interconnect lines and transistors which were originally insulated by the SiO_2.

Chih-Tang Sah is the Pittman Eminent Scholar Chair and a Graduate Research Professor at the University of Florida since 1988. He was a Professor of Physics and Professor of Electrical and Computer Engineering, emeritus, at the University of Illinois at Urbana-Champaign where he taught for twenty-six years and guided 40 students to the Ph.D. degree in electrical engineering and in physics. He has published about 250 journal articles and given 100 invited lectures in China, Europe, Japan, Taiwan and the United States on transistor physics, technology, and evolution. He received two B.S. degrees in 1953, in Electrical Engineering and Engineering Physics, from the University of Illinois, and the M.S. and Ph.D. degrees from Stanford in 1956. His doctoral thesis research was on traveling-wave tubes under the tutelage of Karl R. Spangenberg. His industrial career in solid-state electronics began with William Shockley in 1956, and continued at the Fairchild Semiconductor Corporation in Palo Alto from 1959 to 1964 until he became a professor of physics and electrical engineering at the University of Illinois in 1963. Under the management of Gordon E. Moore, Victor H. Grinich, and Robert N. Noyce at Fairchild, Sah directed a 65-member team on the development of the first generation silicon bipolar and MOS integrated circuit technology including oxide masking for impurity diffusion, stable Si MOS transistor, the CMOS circuit, origin of the low-frequency noise, the MOS transistor model used in the first circuit simulator, thin film integrated resistance, and Si epitaxy process for bipolar integrated circuit production. For contributions in transistor physics and technology, he received the Browder J. Thompson best paper prize for an author under thirty, the J. J. Ebers and Jack Morton Awards in Electron Devices all from the IEEE, the Franklin Institute Certificate of Merit, the first Achievement Award in High Technology from the Asian American Manufacturer Association, and the Doctor Honoris Causa degree from the University of Leuven, Belgium. He was listed in a survey by the Institute of Scientific Information as one of the world's 1000 most cited scientists during 1965–1978. He is a life fellow of the American Physical Society, the IEEE, the Franklin Institute, a fellow of the American Association of Advancement of Science, and a member of the U.S. National Academy of Engineering.